EVER FEEL LIKE YOU'RE LIVING IN A REALITY
TV SHOW? IT'S BECAUSE YOU ARE.

To my lovely wife, Lisa, who not only saved my life both figuratively
and literally but also gave me the support I needed to
write this and tolerated my endless pounding on a
manual keyboard. It can be quite noisy!

DON'T WATCH THIS

HOW THE MEDIA ARE DESTROYING YOUR LIFE

MICHAEL ROSENBLUM
AUTHOR OF *iPHONE MILLIONAIRE*

Skyhorse Publishing

Skyhorse Publishing books may be purchased in bulk at special discounts
for sales promotion, corporate gifts, fund-raising, or educational
purposes. Special editions can also be created to specifications. For details,
contact the Special Sales Department, Skyhorse Publishing, 307 West 36th
Street, 11th Floor, New York, NY 10018 or info@skyhorsepublishing.com.

Skyhorse® and Skyhorse Publishing® are registered trademarks of Skyhorse
Publishing, Inc.®, a Delaware corporation.

Visit our website at www.skyhorsepublishing.com.

10 9 8 7 6 5 4 3 2 1

Library of Congress Cataloging-in-Publication Data is available on file.

Cover design by Brian Peterson

Print ISBN: 978-1-5107-5827-8
Ebook ISBN: 978-1-5107-5876-6

Printed in the United States of America.

Contents

Contents

Introduction

THE FIRE THAT BURNS WITHIN US

At the age of 50, I completely changed my life.

I was a recently divorced, very highly paid executive in the TV business. I lived in Midtown Manhattan in a large glass-and-steel building that towered over Rockefeller Center. My world was defined by meetings in other glass-and-steel towers, by limos and town cars, and by expensive restaurants and dinner parties in other apartments not all that different from mine, with people not all that different from me. I thought I was at the pinnacle of life, that things could not possibly be better. Who would want to live in any other way?

Then, shortly before my fiftieth birthday, I picked up BBC as a client, and I started to commute to Britain on a regular basis. And it was there, at the BBC, that I met Lisa Lambden. She was running the project that I was working on for the Beeb.

I went on to marry Lisa. We bought a small cottage in a little village in the English countryside, and that is where this story properly begins.

I have always been an early riser. Even if I go to bed at two in the morning, I am still up at five. I can't help it. My father was in the army, and he would come into my room every morning at five, turn on the lights, make the sounds of the bugle call of reveille, sans bugle, and then shout, "Time to get up. *Up, up, up*! Out of the sack. Get those feet on the floor!" And that is how I started every day from the age of five or six. Those early lessons live on, so to this day, my eyes pop open at five, and there is nothing I can do about it.

On one particular morning, I was wide awake at five, lying in bed, when I heard a peculiar noise just outside my window—a kind of gentle cooing, but quite loud and quite close.

It was a pigeon, a wood pigeon, and he was strutting back and forth on the peaked roof of our garden room, just below our bedroom window. He had a twig in his beak, and he seemed to be scoping out the neighborhood, perhaps for the local cat that always prowls the garden.

I watched in silence, as I might watch a Discovery or National Geographic nature show, and, after a rather interminable five or six minutes—at least in the world of television where everything happens quite quickly—the pigeon, carefully looking both ways to make sure all was clear, plunged headlong into a tall hinoki cypress tree that stands just at the edge of the garden room.

A few minutes later, the bird emerged from the dense and nearly impenetrable center of the tree, paused on the garden-room roof, surveyed the surroundings, and took flight. A few more minutes later, the pigeon returned to the roof, another twig in beak, and again, after a suitable pause and examination of the surroundings, plunged yet again into the tree.

I was able to watch this show every morning, day after day, for several weeks. The bird was busy building a nest. Soon, I assumed, there would be a whole family of new wood pigeons in my garden.

The tenacity of the birds, their utter and unflinching dedication to the task of building a nest, first from twigs and later from bits of string or long grass that they found God only knows where, astonished me.

How do they know what to do? I wondered. *Who taught them how to make a nest?* A quick visit to the website for the RSPB, or the Royal Society for the Protection of Birds, a most British institution, quickly answered my question. It is instinct. It is in their genes. They not only make nests, they make nests that are unique to their species.

Generation after generation, year after year, these birds would fly off somewhere, find a suitable place safe from cats and other predators, and build a nest identical to the one in which they had been born. It was their instinct that had allowed the species to survive. Without that, they would have vanished long ago.

The longer I watched, the more the question gnawed at me. How did they know? How did they know what twigs to gather, what grass to get? How did they know that spiders would not only feed their young but also provide sticky webs that would help hold their twigs and grass and string bits together like a kind of cement? Who told them that?

The answer, of course, is that no one told them a thing. No one had to teach them. They knew because of instinct. It is in their genes from the moment they are born.

Recently, we went to a small agricultural exhibition at Chatsworth, the massive stately home owned by the duke and duchess of Devonshire, but, as with most of these massive stately homes, opened to the paying public. Apparently dukedom does not pay the way it used to.

Walking the grounds of the agricultural exhibit, we came upon a donkey that had just given birth. The young foal was no more than a few hours old, and yet it was already up and walking and nosing at the hay on the ground in search of something to eat. A collection at the chicken house showed the same thing; young chicks, just hatched, were already busy pecking at the ground in search of food. Everywhere I looked, whether it was to birds or donkeys or horses or the innumerable sheep that live next to my cottage, or even insects— all living creatures seem to have an inherent and inborn instinct that tells them what to do.

My neighbor Tim is a beekeeper. Generation after generation of his bees continue to forage in the field, find their way to the sweetest nectar, and return to the hive to produce wax and honey and yet more bees. Their instincts provide them the knowledge they need for survival.

Conversely, take a human baby, just born; place it alone on the ground; and see what happens. The answer is, of course, nothing. Left alone, it will die.

We humans are an incredibly weak species—fragile, delicate, and seemingly lacking in even the most basic instincts for survival that every other living thing seems to naturally possess. Why is that? How have we not only survived but also risen to dominate the planet and all other living things?

I am not the first person to ask these questions. They are as old as humanity. The ancient Greeks dealt with these very questions, in their own way, nearly three thousand years ago. In the the eighth century BCE, the Greek poet Hesiod tells the story of Prometheus to help explain this dilemma.

Prometheus and his brother Epimetheus were Titans, forerunners of the gods. They were given the task by Zeus, king of the gods, to go to earth and create from clay all living things. Epimetheus went first, creating all the animals of the land and creatures in the sea. Because he finished first, he was able to endow his creations with the best possible gifts for their survival: claws and beaks, razor-sharp teeth and talons, speed and cunning, enormous strength and the ability to fly, and, no doubt—though Hesiod does not mention it—instinct. He took them all.

When poor Prometheus was finished with his creation, man, there were no gifts left to offer. Man stood naked and frail and alone, just like that baby. Prometheus realized that his fragile creation had no chance of surviving even one day in a world filled with Epimetheus's powerful and predatory creations. Mankind would be wiped out in an instant.

Prometheus went to Zeus and asked him if he might give mankind the most powerful thing that the gods possessed, and that was fire. But Zeus was protective of fire. Fire was far too powerful to give to man. With fire, man could forge metals, make weapons, and perhaps one day become so powerful that he would not only attain dominance over all animals but might even challenge the gods themselves. Zeus said no.

So Prometheus instead snuck up to Mount Olympus, home of the gods, and stole fire and gave it to man, his creation, so that mankind might survive.

Zeus was greatly angered. For his transgression, for stealing the gift of fire and giving it to mankind, Zeus ordered that Prometheus be seized and chained to a rock in the Caucasus, where each day an eagle, symbol of Zeus, would tear out and eat his liver, and each night, the liver would regenerate, only to be torn out and eaten again the next day. Prometheus would linger there, chained to the rock, until he was ultimately freed by Hercules, many years later.

The story of Prometheus, is, of course, a story—a myth, a legend. But it was a story with a very specific purpose. It helped the ancient Greeks answer questions that they otherwise had no answers to, like "How did we get here?" or "How do we survive?" Those stories helped them explain a world that was otherwise overwhelming and incomprehensible to them. But the stories did more: they were also instructional.

Human beings, unlike pigeons, are born without natural instincts. We must continually educate each successive generation in every aspect of what is necessary to survive a dangerous world. We tell stories. This is how we pass knowledge that is necessary for survival on to our children, and their children after them, ad infinitum. These stories are far more than entertainment. They are the very tools of our survival.

There is something almost magical about the power these stories have. We remember them, and we repeat them over and over. Ask an average person on the street who Prometheus is, and most people will know, even though he is a character from a story nearly three thousand years old. He is still alive in our cultural mind. How is that possible?

We may not innately know how to build a house or hunt for animals or make shoes or build a fire to keep warm or smelt metal. We aren't born with an instinctive knowledge of how to harvest grain or mill it. We don't know how to fish or gather berries and nuts or know what plants are healthy and which are deadly. We did not get the claws, the razor-sharp teeth, the hard exoskeletons, the horns, the speed. All survival skills, and so much more, so much learned and accrued knowledge over thousands of years of trial and error, life and death, had to be passed on to each successive generation, or they would be lost forever. We did this through our stories. It was this gift of storytelling, more than anything else, that allowed us as a species not just to survive but to thrive.

The gift of Prometheus, then, was not really fire. It was the ability to tell, remember, and then retell a story. It was being able to pass on knowledge from one generation to the next. Without that unique human characteristic of storytelling that no other species on earth has, we would have vanished a long time ago.

If the ancient Hebrews wanted to instruct their followers in the basic precepts of living a good life, they could simply have issued the Ten Commandments as that—ten good rules to live by. However, stripped of the compelling narrative that accompanies the Ten Commandments, the story loses its power. The book of Exodus is a great example of the extraordinary power of storytelling and the everlasting grip that it has on human beings. It clearly demonstrates the way that a highly personalized story, filled with characters and adventures, captures our imagination in a way that a mere recitation of facts or laws simply does not.

In 1949, Joseph Campbell, author and educator, published the seminal guide to storytelling, *The Hero with A Thousand Faces*. In it, Campbell, a professor of literature who wrote about comparative mythology, explained that almost every story that has resonated in almost every human culture, from Greek myths to the present, has followed a basic arc of story he referred to as the monomyth. Whether it was the story of Odysseus or Moses or Luke Skywalker, the plot and the story remain the same, regardless of the specifics. Campbell postulated, and I think proved quite successfully, that we have, in essence, been telling the same story, over and over again, for more than three thousand years. There is, Campbell believed, citing Freud often throughout the book, something very deep in our basic makeup that resonates with this singular structure of telling a story.

The Book of Exodus follows the Campbell model quite well. Moses, an Egyptian prince,[1] is called upon by God (unwilling average person faced with a superhuman challenge), to take the Jews from

---

1 For a fascinating analysis of why Moses might not have been a Hebrew at all but rather an Egyptian prince caught in a struggle of which Egyptian god to worship, read Freud's *Moses and Monotheism*.

slavery into the promised land. This story resonates so well that now, some 3,500 years after the fact, people still know it, movies are made about it, Jews still celebrate it every year as Passover, Jesus had his Last Supper over it, and Zionists will still say, "God gave us this land." That's the astonishing power of a story to educate.

You will note, once you start to look at it, that the Moses story is in fact little different from the Odyssey, from the Jesus story, from the Muhammad story, from the Buddha story. They are all stories that have an enormous impact on our culture, our history, and our society. Stories resonate, educate, shape whole civilizations, and define the daily lives of billions of people. We don't have DNA-inbred instincts like the pigeons outside my window. We have, instead, a very basic, DNA-inbred intuitive hunger for stories.

This addiction to hearing the same story, over and over again, served us extremely well as a species for thousands of years. Our innate attraction to storytelling and the lessons that it conveyed, was great when we listened to perhaps one story a week or so—perhaps even fewer. How often do the Jews repeat the story of Moses and the Exodus? Once a year. In ancient Greece or Rome, how often did your average citizen attend a performance of Aeschylus or Plautus? A few times a year, perhaps, at best. In Christianity, the story of Jesus, from miraculous birth in Bethlehem to his crucifixion and resurrection, makes up the locus of our year, from Christmas to Easter and back again. It is no different from the Jews, who each Saturday read from the Torah, passage by passage, until, over the course of a single year, they reach the end, and then, at the holiday of Simchat Torah, they read the last passages of the book of Deuteronomy and the first passages of the book of Genesis—the end and the beginning, the alpha and the omega. In the Islamic world, it is the power of the story of the prophet Muhammad that is crystalized in the Koran and is recreated each year at the Haj in Mecca. In Elizabethan England, how often did one attend a performance of a Shakespeare play? Also a few times a year? And yet Shakespeare's plays were history writ large, conveying lessons of human nature and human frailty. They too still resonate with us today.

Thus was the human psyche both stimulated by storytelling and inherently limited in the amount of storytelling we could be exposed to. Then, at the very end of the twentieth century, new technologies began to make access to compelling storytelling—done in ways never imagined before, and far more captivating—more and more available.

The invention of moving pictures in the latter 1800s and the ongoing maturation of that technology meant that highly stimulating visual storytelling was now available cheaply and with few complications. But even in the 1930s, a family might go to the movies once a week for a total of about an hour a week of movie-viewing—a lot more than in the past, but still manageable.

Then, in the 1950s, something terrible happened. What had been the wellspring of our ability to survive suddenly turned on us, and, like a cancer, began to threaten our very existence. Like an addiction to any drug, the drive for ever more and ever better visual stories found itself a home, first with the advent of television and then online. Today, the average person can spend an extraordinary eight hours a day or more watching stories in the form of movies, TV, or online video of some kind. What had once been an occasional exposure to a bit of sugar or morphine is now a gluttonous addiction, a mass orgy of never-ending stimulation and information being thrown at us continually.

What happens to a society that spends all of its time watching stories? What happens to a culture endlessly entertained, over and over again, hour after hour, day after day? What happens to a world addicted to endless watching? Our innate hunger for being told stories, once the key to our survival, has now become a helpless and hopeless addiction to a technology-driven, never-ending supply of limitless entertainment, all available at the touch of a button.

So, wired to want to hear more stories, we just keep pushing that button, over and over and over and over. We can't stop. And those stories are not just passive entertainment. Because of our thousands of years of experience, those stories are still teaching us something. We are still busy incorporating their lessons. But what exactly are they teaching us?

There is a school of thought among archaeologists that the Roman Empire collapsed because the Roman elite had inadvertently destroyed their own minds. Almost all homes of the well-to-do had water piped into them. They had remarkably advanced plumbing, the likes of which would not be seen in Europe for another 1,500 years. But their plumbing had a flaw: their pipes were made of lead, and so it is possible that, over time, the Romans slowly but surely poisoned themselves. Lead poisoning leads to slow but certain destruction of the brain.

We don't have lead pipes bringing water into our homes any longer, but we do have fiber-optic pipes bringing content. And it is entirely possible that we too are poisoning our minds, just with a different poison.

In the next two hundred pages, you'll learn how our ancient dependence on storytelling has morphed into a deadly addiction to an almost never-ending stream of entertainment that has warped our world, our society, and our individual lives in ways we never imagined possible. But you'll also learn how you can take back control of your world and your life using the very same media that is destroying us.

PART I

⊘ ⊘ ⊘

THE DRUG

To the Moon, Alice

At 10:56 p.m. Eastern Daylight Saving Time, July 21, 1969, American astronaut Neil Armstrong opened the door of the Lunar Lander, turned around, and slowly began to descend the steps to the surface of the moon.

Some 240,000 miles away, nearly one billion people followed his progress—six hundred thousand of them watching on TV. It was the largest television audience in history, and with good reason. All over the world, people were watching the culmination not just of ten years of intense technological and scientific effort but also ten thousand years of dreaming and imagining a human being walking on the moon.

As it happened, at the very moment that Armstrong was taking his first steps on the lunar surface, something else was happening in space as well. Only a few weeks before the Apollo landing on the moon, the Starship Enterprise, part of the United Federation of Planets, had received a distress signal from the planet Camus II, the site of an important archaeological excavation. An interplanetary research vessel had crash-landed there, and the two survivors were Dr. Janice Lester, a woman with whom Captain James Kirk, commander of the Enterprise, had once had a relationship, and Dr. Arthur Coleman, a famous space archaeologist and scientist.

Almost as soon as Captain Kirk and his crew set foot on Camus

II—almost at the same moment, ironically, that Neil Armstrong and Buzz Aldrin were setting foot on the moon—things began to go wrong, not on the moon, but on Camus II. By the use of an advanced piece of alien technology, one known only on Camus II, Dr. Lester was placed inside Captain Kirk's body, and Kirk was placed in Lester's. And so it was Dr. Janice Lester, in the guise of Captain Kirk, who was now in control of the Starship Enterprise. Would the crew of the Enterprise be able to figure it out before Captain Kirk, who was actually Dr. Lester, was able to carry out the death sentence that the hastily ordered court-martial had handed down to Commander Spock for mutiny?

Back on the moon, Armstrong was busy negotiating the ladder.

Meanwhile, a fleet of Klingon warships was fast approaching Camus II. Armed to the teeth with sophisticated weapons that could easily obliterate the Enterprise and her crew, time was of the essence.

Back on the moon, astronauts Armstrong and Aldrin were busy collecting rocks.

Both the adventures of Commander Neil Armstrong and of Captain James T. Kirk were delivered to Americans though the same exact medium: a screen in their living rooms. One adventure was real, the other fiction. Did it matter? Could anyone actually tell the difference?

Without realizing it, Americans in July of 1969 suddenly found themselves at the crossroads of their future. Would they continue to inhabit the real world of NASA and actual moon landings, boring though they might be, or would they opt to live in a far more exciting world of images, illusion, and fantasy? The lunar landing was the culmination of ten thousand years of grappling with the hard work necessary to accomplish real achievements. The trip to Camus II was the product of just the past few decades, a time of the unleashing of astonishing new technologies used to create and distribute visual content. A great deal of money had been spent to put astronauts in space, but, ironically, vastly more had been spent, in a sense, to put Captain Kirk on Camus II. Spaceflight technology was expensive, but focused—a one-time affair with an ending. Mass-media technology, on the other hand, was a never-ending global enterprise. Neil Armstrong's first step would be a step into the future, but what kind of a future would it be? Would it be a future of real things, or would it be a future of invented realities?

Space exploration and television had, in fact, been born together. They had grown up almost hand in hand. At the height of the Cold War, on October 4, 1957, the Soviets had taken a mighty technological leap over the Americans in what would come to be dubbed the "Space Race." They had launched Sputnik, the first man-made satellite to circle the earth. By this time, television had pretty well penetrated the American market. Almost every household in the country had a TV in their living room, and the news of Sputnik both captivated and terrified the nation.

Once the newspapers, then radio, delivered news like Sputnik, but by the late 1950s, television had begun to eclipse them to become our number-one medium of information. We began to experience the world through images on TV screens.

If anyone was a product of America's embrace of television, it was John F. Kennedy. In 1960, the young junior senator from Massachusetts was making a run for the presidency. He had, at the start, been considered something of a long shot. He was only forty-three years old, had almost no prior experience (he had only been in the Senate for a single and rather unremarkable term), and was a practicing Catholic, considered to be a great handicap at that time.

Richard Nixon, the Republican candidate, by contrast, came to the election as perhaps the most qualified candidate in many years. He had had a stellar congressional history, having cast crucial votes on some of the most important legislation at the time, as well as having single-handedly unmasked Alger Hiss's alleged dealings with the Communists—a very hot topic in the 1950s. Only one year earlier, Fidel Castro, a Communist revolutionary, had overthrown the government of Cuba and had set up a Communist regime only ninety miles from Miami in Florida. Communism seemed to be on the move, and the threat of a Soviet ally just off the coast of the United States was the singular focus of people's attention in 1960, as the election drew close. More than anything else, Americans were searching for a president who could deal with the suddenly quite ominous Communist threat.

That would seem to have been Dick Nixon. He had, after all, been named and elected to the vice presidency of the United States under the very popular Eisenhower and had served for eight years on his reputation as a Communism fighter—and that was before Castro.

In the newspapers, the then-dominant medium of the day, Nixon was the unquestioned front-runner for the presidency. The TV debate, however, would change all of that in an instant. It was neither the content of the debate, nor what was said. There was no gaffe by either candidate, there were no glaring errors, and there was no real singular attack. Television would change the election not because of policy, but rather because of perception. One candidate was great on TV; the other, terrible.

Kennedy arrived for the debate at the Chicago studios of CBS on September 26. He was well rested and well tanned. He looked good. Kennedy had spent the entire prior weekend holed up in a hotel room, practicing debate questions and answers with this team.

Nixon, on the other hand, arrived at the debate directly after a grueling week on the campaign trail. It had been a bad week for Nixon all around. He had been suffering from the flu. He had lost nearly twenty pounds, and his suit hung on him like a sack. On top of that, he banged his knee getting out of the limo when he arrived at CBS, exacerbating an earlier injury. He was in pain, gray, and sallow. In short, he was bad TV.

Despite Nixon's lack of preparation and difficult week, he knew the material cold. He had lived it for years. He was, in fact, the very embodiment of US anti-Communist policy and action. For those who listened to the debate on the radio, Nixon was the narrow winner, but for those who watched the debate on TV, Kennedy had buried Nixon. It was not about substance any longer; it was about who looked better.

Following the debate, Richard J. Daley, the Democratic mayor of Chicago said, "My God, they've embalmed him [Nixon] before he even died." The next day, the *Chicago Daily News*, a Republican-leaning tabloid, ran the headline, "Was Nixon Sabotaged By TV Makeup Artists?"

As it turned out, neither Nixon nor JFK had taken the option for makeup artists, which CBS had offered. But Kennedy was well tanned and looking good. Nixon, who had always suffered from a five-o'clock shadow, simply came across as shifty and sinister.

Nixon had once confided to CBS anchor Walter Cronkite, "I can shave within thirty seconds before I go on television and still have a beard." Concerned about this, he had, at the last minute before the cameras started to roll, conceded to his support staff who suggested that he at least apply a thin coating of a product called Lazy Shave, a kind-of drugstore over-the-counter pancake. Nixon had used this many times before while working a crowd, and it had been fine. But under the hot studio lights, Nixon, who had a tendency to sweat anyway, began to perspire and the Lazy Shave began to melt off on screen. It was a visual disaster. As time went by, he began to look like a zombie in a B-movie.

But even before the debate, Kennedy had understood the power of visual media. He had invited documentary filmmakers Albert Maysles and Richard Leacock to follow him as he worked the state of Wisconsin prior to his primary run against Hubert Humphrey. Using small, hand-held cameras, a cutting-edge technology in 1960, Maysles and Leacock produced a groundbreaking film called *Primary*. It presented the candidate in an entirely new and, until then, unseen light. Maysles and Leacock created an intimate and highly personalized story of what it was like to run for the presidency. JFK was cast, and there is no other word for it, as a character in a movie with a great story arc. He went from being a political figure to a personality.

Today, we accept this as the norm, but in 1960, no one had ever done this before. Political stump speeches, the bread and butter of electoral politics up to that point, are, frankly, boring. Stories about underdogs who work tirelessly to try to win make great movies. JFK was the "Rocky" of Wisconsin. He went on to win, but more important, he had begun the great transformation of politics and its relationship to the visual media.

If *Primary* was the harbinger of the coming power of television and what visual storytelling would mean in politics, the great television debate between Kennedy and Nixon could seal the change. Kennedy had gone into the TV debate virtually tied with Nixon. By

the time the four debates were over, he was the narrow front-runner. Kennedy, the better TV personality, went on to win the White
House. In 1962, Nixon would publish his memoir, *Six Crises*, in
which he would acknowledge what had happened. "I should have
remembered that a picture is worth a thousand words," he would
write. As it turned out, a picture was worth a good deal more than a
thousand words. It was worth the White House, the presidency, and
the nation.

On April 12, 1961, a mere 90 days after Kennedy had been sworn
in as President of the United States, a Vostok-K rocket blasted off
from the Baikonur Cosmodrome in Southern Kazakhstan, USSR.
Atop the rocket was the Vostok-1 space vehicle, and inside the small
vehicle was Soviet Cosmonaut Yuri Gagarin.

Seven minutes after blast-off, Gagarin became the first human
being to orbit the earth in space. One orbit and 108 minutes later,
Gagarin and the Vostok landed back on earth, bouncing off the
ground, eighteen miles south of the city of Engles in the Sarotov region
of the Soviet Union, 170 miles west of the intended landing site at
Baikonur. Seven kilometers above the earth, as Vostok-1 descended,
Gagarin opened the hatch and was ejected from the spacecraft. He
glided to earth, as planned, landing separately in a field.

A farmer and his daughter came upon Gagarin, gathering up his
parachute.

"When they saw me in my space suit," Gagarin would report later,
"and the parachute dragging alongside as I walked, they started to
back away in fear. I told them, 'Don't be afraid, I am a Soviet citizen
like you, who has descended from space, and I must find a telephone
to call Moscow!'"

The news, of course, was filled with this astonishing Soviet
achievement. Americans felt that they were being left far behind. The
Space Race had begun.

One month after the Russians put Gagarin into the first manned
orbit, Kennedy delivered America's answer to the Soviet Union. In a
speech before Congress, Kennedy declared that the country "should
commit itself to achieving the goal, before this decade is out, of

landing a man on the moon and returning him safely to the Earth."
This was, at that moment, an absolutely astonishing thing to suggest.
When Kennedy laid down the challenge of a flight to the moon, the
US space program, such as it was, had only put one astronaut, Alan
B. Shepard, into a suborbital flight—essentially up and down, just
three weeks earlier. The whole thing had lasted a total of five minutes
and twenty-two seconds. Now, to say that we would go to the moon,
land, and come back in essentially nine years must have seemed the
height of insanity. Many people told him it was both ridiculous and
impossible.

In a later speech at Rice University, Kennedy neatly outlined the
obstacles that lay before them, should they accept the challenge:

> But if I were to say, my fellow citizens, that we shall send to the
> moon, 240,000 miles away from the control station in Houston,
> a giant rocket more than 300 feet tall, the length of this football
> field, made of new metal alloys, some of which have not yet been
> invented, capable of standing heat and stresses several times more
> than have ever been experienced, fitted together with a precision
> better than the finest watch, carrying all the equipment needed
> for propulsion, guidance, control, communications, food and
> survival, on an untried mission, to an unknown celestial body,
> and then return it safely to Earth, re-entering the atmosphere at
> speeds of over 25,000 miles per hour, causing heat about half
> that of the temperature of the sun—almost as hot as it is here
> today—and do all this, and do it right, and do it first before this
> decade is out—then we must be bold.

The challenge JFK outlined had all the hallmarks of a Joseph
Campbell impossible mission that had to be accomplished. It was
the purest essence of great storytelling, and, as such, it captured the
nation's attention. Married to the astronauts, who were cast as "the
right stuff" and heroes for our time, it created an almost perfect
Campbellian construct of an epic tale. Also in the speech at Rice
University, he said:

"We choose to go to the moon! We choose to go to the moon in this decade and do the other things, not because they are easy, but because they are hard, because that goal will serve to organize and measure the best of our energies and skills, because that challenge is one that we are willing to accept, one we are unwilling to postpone, and one we intend to win, and the others, too."

This is how you inspire a nation to do the impossible—with a story. The power of storytelling, married to television and visuals, is almost beyond belief. When you look at the Apollo program, the application of character and story to real financial backing, it frankly is beyond belief—but it happened.

Kennedy, the nation's first television president, had committed the nation to achieving the inconceivable and almost impossible task. Yet a mere seven years after the Rice University speech, America had accomplished the impossible. Neil Armstrong walked on the surface of the moon.

Placing a man on the moon was by any measure an enormous achievement. It was more than the culmination of a promise made only seven years prior. It was, in many ways, the fulfillment of a dream that had captured people's imaginations since they first learned to walk erect and look up at the sky. The moon had always been a tantalizing object. It was close that you could almost hold it in your hands, yet it was so very untouchable—until Neil Armstrong put his foot down on the lunar soil.

The completed dream of leaving the bonds of the earth and touching the moon—of walking on it—felt, in those heady moments, as though we had all, as a species, crossed some kind of historical line. We would no longer be bound to the earth. We would begin the next great chapter of humanity, the exploration and ultimately the colonization of the moon, of Mars and all else that lay beyond. It was a remarkable moment.

Yet only a few short years later, on December 14, 1972, American astronaut Eugene Cernan climbed into the lunar module, shut the door, fired up the engine, and left the surface of the moon. He was the last man to walk on the moon, a mere three years after Neil Armstrong had been the first. Armstrong had said on setting foot on the moon, "That's one small step for man, one giant leap for mankind"; Cernan's last words as he stepped into the Lander were, "And, as we leave the moon at Taurus-Littrow, we leave as we came and, God willing, as we shall return."

But no one would return, ever again, to this day. Why? What killed our passion for exploration and the unknown in such a short time, a passion that had once been so strong that we accomplished the seemingly impossible?

It wasn't a matter of the cost. The original Apollo lunar program had cost a total of $25.4 billion in 1969 or about $180 billion in current money.[2] That was spread over ten years, so even in current dollars, that would come out to a commitment of about $18 billion a year. Now, to put this into perspective, the current budget of the Department of Defense for just one year was $639.1 billion. Even so, the vast majority of that $25.4 billion was spent in developing entirely new and hitherto untested technologies, processes, and machinery. Think of it as sunk cost. Think of it as investment in the future, a future that was cut short. The specific trips to the moon and back cost only a fraction of that. And even that is a pittance when compared to the $5 trillion or so that has been spent on the wars in Iraq and Afghanistan. So, if we wanted to go, we could certainly have afforded to go, any time we liked. But we did not want to go.

Nor was it a matter of the technology. The technology that took men to the moon at the end of the 1960s was like your grandfather's 1969 Chevrolet—cutting edge in the day, but now really museum-grade stuff. Your iPhone, for example, can perform functions 120 million times faster than the IBM mainframes that guided the Apollo 11 to the moon. The Apollo module had 2K of memory and 32K of storage. Your toaster most likely has more computing power than the lunar module. The original Apollo program was supposed to go all the way to Apollo 20. Instead it got axed after Apollo 17. Why? Once we finally got to the moon, no one got killed. No one

_____

2 "1969 Moon Landing," History Channel website, https://www.history.com/topics /space-exploration/moon-landing-1969.

even got injured. Even with the exception of the notorious Apollo 13, which made a great movie, everything went off flawlessly.

Like Armstrong, Shepard, and the rest of the astronauts, Gene Roddenberry had also been a pilot in the Second World War. He had flown B-17s over Europe. And when the war was over, like many pilots, he got himself a job working for Pan Am, the commercial airline. Following the war, commercial flying was just starting to take off. Roddenberry was so good a pilot that in a short time he was flying the longest routes that Pan Am had then, New York to Johannesburg and New York to Calcutta. Even though he enjoyed flying and was quite good at it (he probably could have been an astronaut), his real passion had always been writing.

On June 18, 1947, Captain Roddenberry got into trouble. Today, plane crashes are rare, but in the 1950s, the technology behind big planes was a good deal less dependable than it is today, and if planes did not crash with regularity, they certainly crashed far more often than they do today. His plane, the Clipper Eclipse, came down in the middle of the Syrian Desert. Roddenberry survived the crash with two broken ribs, and even though he was injured, he was able to drag passengers out of the flaming wreck. Fourteen people died in that crash, and for Roddenberry, it was his last commercial flight. He resigned from Pan Am as soon as he got back and decided to finally pursue his long-time passion for writing. Little did he know that he would also fly into space, just like his companion pilots at NASA, only in a very different way.

Roddenberry took a rather circuitous route to his writing career. He joined the Los Angeles Police Department, not as a patrolman but as a public information officer. He ultimately became the chief speechwriter for the LAPD's chief of police. His work at the LAPD led him to become the technical advisor on a TV series called *Mr. District Attorney*. In that capacity, he began to write TV scripts in earnest, selling a few story ideas. By 1956, he felt confident enough about his writing abilities to quit his job with the LAPD and concentrate on writing full time.

Like all writers, he started slowly, but over time, as his abilities and contacts grew, his work become more and more well-known. He started writing scripts for existing TV series such as *Boots and Saddles* and *Jefferson Drum*. Then, in 1958, a script that he wrote for the TV series *Have Gun—Will Travel* won him the prestigious Writer's Guild of America Award for Best Western Episode.

In 1961, he got an idea for a series based on a multiethnic crew aboard an airship that traveled the world, but he was told that there was little market in the TV world for science fiction at that moment. The Space Race would change all of that. On March 11, 1964, he wrote up a sixteen-page treatment for a new series he called *Star Trek*. As with many Hollywood writers, he offered the concept to a number of networks and studios. Most turned him down. Finally, NBC agreed to finance a pilot. It was not so much science fiction as a kind of *Wagon Train* to the stars. The real space program was in high gear, and all of the networks were suddenly looking for some kind of space-related series. CBS commissioned *Lost in Space*, so NBC needed its own space drama. On March 24, 1965, the first episode of *Star Trek* went into production. The pilot was so good that NBC commissioned a full season run of thirteen episodes.

The same week that *Star Trek* went into production on the ground, astronauts Gus Grissom and John Young went into space for real, making the first Gemini space flight. They went for three orbits; the whole mission lasted just over four hours. The USS Enterprise, by contrast, was on a five-year mission, not a three-hour one.

Star Trek and the NASA space program ran concurrently on TV during the 1960s. They were kind of competitors. Like Captain Roddenberry, NASA also went into the TV business along with going into space. NASA had always understood that good television coverage of its missions was essential to keeping the agency funded. An audience that watched the space shots and fell in love with the astronauts and their families would mean that Congress, who voted the funding for NASA, would be responsive to the feelings and passions of their constituencies. Thus, NASA made every effort to make the Space Race into primetime TV viewing.

I am old enough to remember when school came to a complete halt on launch days, and a black-and-white TV set would be rolled into the classroom so that we could watch the launch of whatever mission was heading off into space. The countdowns were great drama: "ten . . . nine . . . eight . . . seven . . . six . . . we have ignition . . . five . . . four . . . three . . . two . . . one . . . blast off!" And the great rocket would rise into the sky as we all held our collective breaths.

It was great TV. In fact, in many ways, NASA invented the world's first reality TV show. When Neil Armstrong set foot on the moon for the first time, NASA made sure that his last steps onto the lunar surface would be beamed back to the largest TV audience ever. This in itself was in many ways as complicated as putting a man on the moon. No one had ever transmitted a live image from the surface of the moon. Everything had to be designed from scratch, including the camera.

By the launch of Apollo 12 on November 14, 1969, just four short months after Armstrong and Aldrin's historic mission, national interest in watching the moon landings had waned considerably, "considering the intense national emotion spent on the first moon landing," the New York Times reported. "'You can't get as excited the second time you kiss the girl,' one man said."

Apollo 12 Commander Alan Bean did not help matters by pointing the mission's only TV camera directly into the sun, burning out the tubes. Left with live audio but no video, CBS cut to a studio on Long Island, where two actors, dressed as astronauts, pantomimed what they assumed was happening on the moon. NBC used astronaut marionettes.

As NASA was losing TV audiences, despite stunts like having an astronaut hit a golf ball on the surface of the moon, Star Trek, the other TV space-travel series, was entering its third season. While NASA TV was desperately trying to hold its audiences with stunts like astronauts eating Jell-O in a weightless environment, Captain Kirk and his crew were cruising the universe at warp-drive speed, fighting Klingons, firing off photon torpedoes, beaming down to planets with green slave women, and outwitting computers that ran entire worlds. It beat weightless Jell-O—by a lot.

Why did we, as an audience, as a nation, invest far more, both financially and emotionally, in Star Trek and Star Wars and their intergalactic adventures than we did in NASA and its real life, real world explorations? It's because Star Trek and Star Wars and the rest were simply more exciting than the real trips to the moon. *Star Trek* involved voyages to far-off galaxies with dangerous Klingons or green slave women. Trips to the moon involved two guys jumping around in grey dust. *Star Trek* involved travel at warp speed, photon torpedoes, phasers on stun. Trips to the moon involved three guys strapped to a bench. *Star Trek* and *Star Wars* were endlessly exciting. Trips to the moon, as with most real science, and most real life as it happens, was often deadly boring.

And because the reality of going to the moon was ultimately visually boring, we all collectively voted to kill the Apollo TV series and extend the *Star Wars/Star Trek* franchises to infinity and beyond. Real trips to the moon made terrible television, and by the 1970s, we were starting to watch a lot of TV all the time.

With bad ratings on TV, the Apollo program lost public support and therefore congressional funding. And because it lost congressional funding, it got killed. American viewers watched the landings on the moon through the same screen that they watched *Star Trek*. For them, in a strange way, there was really no difference between the two. They were both people going into space. They were both offered as a kind of entertainment. The fact that one was real and one was false in the long run ultimately made no difference whatsoever. People are naturally attracted to the more entertaining event. And *Star Trek* proved far more entertaining than real life. So *Star Trek* won. Real life lost.

And, of course, having laid out the parameters of the story in 1961 that we would land a man on the moon and return him, and having accomplished that, the story was over. No one was really interested in how Odysseus' life worked out a few years after he came back to Ithaca, were they?

Since 99.9999 percent of the people in America were never going to go to the moon anyway, what difference did it really make to them if the space shows that they were watching were fact or fiction? When your whole experience of real life is perceived through a screen, you very quickly lose the ability to differentiate between reality and fiction. If I ask you about the Titanic, you immediately have an image of the ship in your mind's eye. You know exactly what it looks like. You have, in fact, seen the Titanic go down. Of course, you haven't really—you've seen the movie, but those images become a kind of reality for you. Dunkirk? You know exactly what it looks like. You've seen it. You've been there. Winston Churchill? Got it. Syrian refugees? Likewise. The War in Vietnam? Know it well! You may be excused if you begin to confuse news footage with *Full Metal Jacket* or *Apocalypse Now*. Korean War? You've seen *M*A*S*H*; you know what it was like. All of these ideas and images, and millions more, were not cast from real life, nor from real life experience. They were the product of TV or movies or video. And this does not stop with movies and TV. The endlessly exciting and beautiful lives that people spend hours a day creating for others on Instagram or Facebook are but a more personalized representation of what Gene Roddenberry did for the banality of space flight in the 1970s. We live increasingly in a world of fiction, either created by Hollywood or now created by ourselves. Most of it is no more real than Captain Kirk's trip to the stars, yet just as compelling and capturing.

It was one thing when people went to the movies once a week, for an hour or so, but quite another when we start to live in a movie theater all of the time. The endless exposure to fiction like *Titanic* or *Dunkirk* creates in our minds a kind of false memory, a false yet very solid certainty that we know what something looked like and felt like, or how something happened. And based on that very strong "memory," we often act.

Today, the average American spends an astonishing eleven hours a day staring at a screen.[3] Our knowledge of what the world is like comes to us through those screens. And what is on those screens very much controls what we think, what we buy, how and where we work, what we believe, for whom we vote, with whom we go to war, and pretty much everything else about our lives and how we live them.

Imagine if television or video had existed in 1492. Imagine if Columbus, making his first exploratory trip to China (or what he thought would be China) by sailing west across the Atlantic had also been equipped with a live TV camera, so that the people of Spain

---

3 Quentin Fottrell, "People Spend Most of Their Waking Hours Staring at Screens," MarketWatch website, August 4, 2018, https://www.marketwatch.com/story/people-are-spending-most-of-their-waking-hours-staring-at-screens-2018-08-01.

could see, in real time, first hand, as he landed on the coast of China! What an exciting event.

Isabella, of course, had been under a lot of pressure once she made the commitment to fund the Columbus expedition. "What an utter waste of money," the vast majority of Spaniards were saying. "All that money could be spent for better things—like a more expanded Inquisition, for example." But her media advisors had told her that "The Mission to China" would make great TV, and that would win over the masses.

On October 12, 1492, Columbus sighted land—China, he believed. "Fire up the video camera," he ordered. And all across Spain, millions of people sat riveted before their TV sets as Columbus landed on Watling Island, in the Bahamas. Ever been to Watling Island? Ever been to the Bahamas? This was before the Paradise Island resort. This was before the One and Only Ocean Club. Pretty much all that was there was sand and scrub.

As an episode, "Columbus Sails the Ocean Blue" was a ratings disaster, right up there with trips to the moon—boring, the worst thing that can befall you in the media world. So *The Christopher Columbus Show* gets cancelled, and no one ever goes back to the New World again.

Fortunately, there was not TV in fifteenth century Spain. Decisions were made more or less on the basis of reality, which in those days took a long time to figure out. The time between event and reportage of the event actually had some benefits. It allowed time to digest information. Today we live in a world of images; we base our judgments and our decisions on images that are communicated to us on a real-time, second-by-second basis on screens, TV, your phone, Instagram, Facebook, Twitter, YouTube, and a million other devices and platforms.

As a result of this, we are increasingly incapable of being able to distinguish between fact and fiction, between reality and imaginary events, between truth and lies. In a world in which fact and fiction comfortably coexist, the line between them begins to erode. Suddenly, anything is possible, and that which is most entertaining, which rates the best often becomes the foundation of belief, and often policy.

Whether it is objectively true or not makes no difference. This is a result of sixty years of getting both our information and our entertainment through addictive, even hypnotic, flashing lights on a screen. And it has consequences.

The result, after just over a half century of nonstop exposure to the medium, is that we now no longer really care, for the most part, if what we are watching is real or fake. It is, to us, all the same. More disturbing perhaps, is that many of us are no longer even capable of differentiating the difference between the two.

We have been fighting a war in Afghanistan for seventeen years now, twice as long as we were in Vietnam, and it seems that there is no end in sight. After a decade in Vietnam, people took to the streets in protest. In America today, we are passive. The images of the war on TV or online are fleeting. More often than not, the only real images one gets of the brutal reality of what is actually happening is when a soldier rotates home and suddenly surprises his or her child at their school—a great TV moment. Do we want to deal with the ugly reality of people actually getting killed? I would say not. That would be bad for ratings.

After the Apollo landings on the moon, we could have gone either way. Stanley Kubrick's 1968 epic film *2001: A Space Odyssey* portended a promising future of Pan Am spaceships and Bell Telephone calls—an extension of the physical world that we were already familiar with.

We stood at the precipice of two futures—one made of fact, the other of fiction, and yet we opted for fiction. In the real world it had taken ten long years to get from the earth to the moon. In the fictionalized world of movies and TV, it would take only another eight years to get from the Apollo landings to Luke Skywalker, the Death Star, and the rest of the universe. In a world of increasingly short attention spans and fast food, the choice was obvious. Did it matter that one was real and the other fiction? Clearly not.

In 2016, Donald Trump, a New York real estate magnate and, more important, the star and executive producer of a top-rated reality TV show, ran for the presidency against Hillary Clinton, former United States senator, former secretary of state, perhaps one of the best prepared candidates in history. Yet Trump, the TV star, would defeat Clinton in a sort-of replay of the 1960 election. Trump was excellent TV; Clinton was terrible. And that made all the difference.

2

The Greatest Story Ever Told:

HOW A REALITY TV STAR BECAME THE MOST POWERFUL MAN ON THE PLANET

> "It [a Trump residency] may not be good for America, but
> it's damn good for CBS."
> —*Les Moonves, former chairman and CEO, CBS*

On September 18, 2016, more than a year after Donald Trump's announcement to run for president of the United States, more than a year after he had gone from national joke to the Republican candidate for president, Jimmy Kimmel was hosting the *68th Primetime Emmy Awards* for television excellence. Kimmel's job was to entertain the assembled.

"Television brings people together, but television can also tear us apart," Kimmel said. "I mean, if it wasn't for television, would Donald Trump be running for president?" It was meant as a joke, but nothing Kimmel would ever say in his career, and nothing he had ever said before, would be so true nor so prescient a commentary on the state of American society.

The media and TV crowd laughed. The *New York Times* was slating Hillary with a 97 percent chance of winning. There was no way in the world that the former host of *The Apprentice* was going into the White House. Perhaps Trump himself did not even believe it.

"Many have asked, 'Who is to blame for Donald Trump?'" Kimmel continued, playing with the crowd who were clearly enjoying the joke. "I'll tell you who, because he's sitting right there. *That* guy."[4]

That guy that Kimmel was referring to was TV producer Mark Burnett. Burnett had made his mark as the creator and executive producer of *Survivor*, one of the very first reality TV show hits. But that night, just weeks before the 2016 Presidential Election, Kimmel was referencing another show, *The Apprentice*, another Burnett production, and specifically its host, Donald Trump.

Burnett had found Trump, not only a mediocre and failing midlevel real estate operator in New York, but also a man who had a burning passion and skill for attracting the attention of the media. He was drawn to the media like a moth to a flame, and they to him as well. Despite his failings or perhaps leveraging off of them and seeing their potential for good ratings driven by their attention, Burnett had crafted a reality show around Trump, casting him as an all-knowing business genius. This was far from the truth, but in the world of reality TV, truth is immaterial. Each week, Trump was shown as a billionaire who knew how to get things done, and for those who failed, well, they were "fired!" in what became the trademark end to each episode.

The show turned into a hit, initially running at the top of the TV ratings, syndicating internationally, and creating a handful of spinoffs, the sure signs of TV success. For tens of millions of Americans who otherwise would never have heard of Donald Trump, he was now well known as one of the most brilliant self-made billionaires in the world (just as Daniel Craig might be thought of as Britain's top spy). Now, the confusion between real life and the world of entertainment fiction was about to be crossed in the race for the most powerful job in the world—the presidency of the United States.

At the Emmy Awards that night, Kimmel would poke fun at Burnett, the man who had created Donald Trump. It would have

---

4 I am deeply indebted to Patrick Radden Keefe, whose excellent article "How Mark Burnett Resurrected Donald Trump as an Icon of American Success" in *New Yorker* magazine provided me with much of this information. See https://www.newyorker.com/magazine/2019/01/07/how-mark-burnett-resurrected-donald-trump-as-an-icon-of-american-success.

been funny, except it was deadly serious. A wholly created reality TV star was on a path to being elected president of the United States.

"Thanks to Mark Burnett, we don't have to watch reality shows anymore, because we're living in one," Kimmel said. Truer words would never be spoken. Not only was a reality TV personality about to enter the White House, but also he would soon bring his ten years of training under Burnett to the entire way he would run the government and the country and deal with the press, the public, and the media. That was as one giant and never-ending race for ratings. Soon we would all be unwilling participants in the greatest or most frightening TV reality show ever imagined.

Nearly thirty-five years ago, Neil Postman published his seminal book *Amusing Ourselves to Death*. It predicted all that would ultimately come to pass. Postman warned that a culture driven by television was inevitably going to be driven by entertainment.

Postman warned that as we learned about the world via television, the only things that would get and hold our attention through that medium would be things that were first and foremost entertaining. We have always been suckers for great stories. For thousands of years, they educated us about what to do. And now we had a medium that was specifically designed to continually deliver compelling stories to millions of people twenty-four hours a day, for free.

Postman wrote at a time when we did not watch all that much television and long before there was an Internet or smartphones that would deliver video to us all the time, no matter where we were. It may be hard for contemporaries to accept the idea that spending hours a day staring at a screen, either on a TV, a computer, a tablet, or now your phone, poses a danger. But it does. It poses a danger to society as a whole; it is also incredibly destructive to individuals. It is the root cause of anxiety, depression, general unhappiness, and far worse.

In 1964, when the Surgeon General Luther Terry released his first report on the dangers of smoking, it was also hard to believe. Almost everyone, after all, smoked, so how dangerous could smoking really be? We used to gorge on fast foods and sugary soft drinks until the nation was hit with an obesity crisis. Could there possibly have been a connection? And now, we have the media.

THE MOST DANGEROUS DRUG IN THE WORLD

Let's start with the basics. The job of the media is to transmit stories to us. It is the consummate storyteller, and we are wired to listen and watch and learn. The media and the stories that they tell us are specifically engineered to be addictive, to plug into our deep desire to hear and absorb exciting and well-engineered stories. They are also directly plugged into our innate connection between stories and education.

Many years ago, I partnered with former Vice President Al Gore, joining up with him shortly after he very narrowly lost the race for the White House. Long before he discovered the dangers of climate change, he had approached me and said he wanted to start a new TV channel that we would own. I asked him why, of all the things he could do, he would want to do that. In his position, a nearly elected president, a former vice president of the United States, a former US senator, he could have done anything, joined any board, been president of a university, started a foundation—why a TV channel?

It was because, he said, starting and owning a cable channel was the fastest way to make money in America.[5] And the reason? Because, he said, TV is addictive. And, as it turns out, he was right. It is a legal drug, like cigarettes. We cannot resist it. Today and every day for the rest of your life you and everyone else will spend between eight and eleven hours staring at screens.[6] All of these are vehicles for delivery of an incredibly addictive drug. And what is the drug? What is the product that the media delivers? Stories.

It seems crazy at first. Who would pay for stories? But that, in fact, is what we crave. Whether the news or HBO or Netflix or Facebook or Instagram, the only product offered is storytelling. What a very strange product to manufacture. You can't eat stories, you can't wear them; you can't use them to warm your home, power your car, or cure your ills. They are almost nonexistent. They have no dimension, no physical presence, no weight, no mass. They are like a dream, a chimera. And yet, the global TV business that produces these gossamer phantoms generates $1.72 billion a year, more than global oil.

5 As it turned out, when we sold Current TV some seven years later for $500 million, that showed he had been right all along.
6 Quentin Fottrell, "People Spend Most of Their Waking Hours."

And that is without any online video, movies, social media, or anything else factored in. Illusions are a massive business, perhaps the biggest business in the world. Steven Spielberg's studio is not called Dreamworks for nothing.

But most Americans now spend a majority of their time staring at screens, enraptured by those amorphous stories. This is an incredible change for society and for the human race as a whole. Until some eighty years ago, most had even seen a screen, let alone spent the majority of a life staring into one. But this drug, this electric heroin, is so incredibly potent that in just one or two generations it has captured not just the bulk of our attention but also that of most of the planet. Like most addictive drugs, someone is making a lot of money out of this, but it isn't you. You are the one paying for the drug—endlessly. This drug is, in fact, bankrupting you.

And what is it that you are spending so much time looking at? Although it is endlessly entertaining, almost without fail, the content of those addictive stories is *other* people doing things. That is the nature of storytelling and the essence of television, movies, and video, and it is the key to its enormous success. We are effectively spending our lives watching other people do things, whether it is a drama, cop show, news program, sitcom, sports event, game show, or program like *House Hunters International.* That is the architecture of the medium: passive watching. When the Internet arrived, its various contents simply began to ape the format and form that movies and television had already cast—watching other people do things that were more interesting and more exciting than our own life. Although the format makes no difference in terms of the passivity, video is dominating. According to Cisco, this year 85 percent of all online content is now video, and that is only going to increase.

Spending your life watching stories, watching other people do things, interestingly creates both a sense of deep frustration and deep depression. Everyone we are watching seems to have a far more interesting life than our own; otherwise, why would we watch them? Watching other people do amazing things and live amazing lives was great when we went to the movies once a week or a play a few times a year. Who could argue with Fred Astaire dancing or John Wayne getting the bad guys? But spend hour upon hour, day after day watching other people endlessly living incredible lives, and you start to realize that your life is far from amazing and will probably always be. The stories that once educated us on how to hunt or how to be a good person are now educating us, over and over, that we are essentially failures.

Both depression and anxiety spiked in this country after the introduction of TV sets in the 1950s,[7] and it has continued to climb upward ever since. Watching stuff may be addictive, but it is also deeply depressing, a condition which leads, sadly, to more watching.

The advent of social media only magnified the problem. Online, we saw a never-ending parade of now-average people as well as celebrities leading amazing lives. The response, for those who post? Increasingly fabricated and mediated stories of our own—our own amazing meals, amazing vacations, amazing parties. Yet in our hearts, we know that this is also a lie, and this only serves to compound the depression and anxiety.

744,600 HOURS

If you live to be eighty-five years old, your entire life will be 744,600 hours long.

If you sleep an average of eight hours a night, your conscious time on this planet, from birth to death is 496,400 hours. This alone is a rather shocking piece of information.

If you spend eleven hours a day staring at a screen, starting, let us say, at the age of five, then you are devoting 321,200 hours of that time to screen-watching. If you are spending eight hours a day watching TV or videos, then you are devoting 233,600 of those hours to watching TV or videos.

Put simply, you will spend about 65 percent of your life staring at a screen and 46 percent of your life watching some iteration of video. You are effectively spending your life watching, living in, and immersed in, a made up, fabricated world. Given the choice, this

---

7 Jeanie Lerche Davis, "Childhood Anxiety Steadily on the Rise Since the 1950s," WebMD, December 14, 2000, https://www.webmd.com/children/news/20001214/childhood-anxiety-on-the-rise#1.

probably is not how you would prefer to spend your life, but that is how you are spending it—you and everyone else. This is the hard truth of our addiction.

It is almost as though, a little more than half a century ago, we all began, without our knowledge or permission, to participate in a kind of massive psychological experiment. In the late 1950s, we all agreed, more or less, to see what happened if we spent eight to eleven hours a day, every day, staring at a screen for the next sixty years. It did not have to be this way. We could just as easily have asked what would happen if we all, collectively, committed to spending eight to eleven hours a day, every day, playing tennis.

Had we all, collectively, in 1955, decided that for the next sixty years or so, we would all of us devote eleven hours a day, every day, 365 days a year playing tennis, we would by now be the most athletically fit nation in human history. We would all wear tennis whites all the time. We would spend billions of tax dollars on tennis courts. We would have a foreign policy devoted to building tennis courts around the world in developing nations. "No Child Without a Racquet" would be a congressional priority that everyone could get behind. We would have TV debates: grass versus clay. People would probably kill each other over it. We would elect the president of the United States based on his or her forehand.

Had we all, collectively, sixty years ago or so, decided that each of us would spend eleven hours a day, every day, for our whole life practicing the piano, we would be the most musically proficient society in human history. We would hum Beethoven and Bach all day long. At the age of three, in order to distract our children, we would plunk them down in front of little tiny grand pianos and let them plonk away. Our cars would be outfitted with little keyboards in the back seats so kids could play Rachmaninoff to distract them on long drives. Composing while driving would be a national automotive issue. Pulled finger tendons would be a national disease, and we would have a "War on Finger Tendonitis" funded by the government. People would go to war over Bösendorfer versus Steinway. We would have piano-concerto competitions to elect our leaders. Instead of debates, we would have "play offs."

We might, in our theoretical what ifs, have decided in the early 1960s that we were all going to commit to reading books for eight to eleven hours a day, every day, for the next sixty years. What a society we might have created!

But, of course, we didn't elect to practice the piano or play tennis or read eleven hours a day, every day for our entire lives to see what would happen. We elected instead to spend eleven hours a day, every day, for our entire lives, and the lives of our children and of our children's children staring at images of other people doing things. That endlessly repeated activity had a marked influence on us. It, more than anything else, made us who we are today.

WHAT DID IT TEACH US?

Devoting the majority of our time to endlessly staring at screens did lots of things to us, both collectively and individually. Non-stop screen watching drove our never-ending need for celebrity, for fame, to acquire things. It created a world in which everything was either entertaining or not worthy of our interest. It made us obese. It destroyed our ability, largely, to focus on the written word. It endlessly flooded our world with compelling images of attractive people living amazing lives, lives that you will never be able to replicate. It is designed to make you feel bad about yourself. It is designed to make you believe that you can fill that hole in your heart with the stuff that the advertisers are offering, even if that is not at all possible. It put a TV reality star in the White House. It gave us a very short attention span.

But beyond all that, it did something else, something far worse in many ways. And it happened when we weren't even looking—or actually, when we were spending all our time looking.

IT MADE US PASSIVE

Watching is, by definition, a passive activity. You are supposed to sit quietly and listen, or watch, when someone tells you a story.

People who play tennis or play the piano are actually doing something. They are, in a way, creating something. Even reading is a far more interactive activity than watching. It engages your brain in a fundamentally different way.

How many times have you heard on TV or on a video "just watch this!"? The emphasis is really on the "just watch" part. How many times have you called up a friend and said, "What are you watching?" as though the act of watching was the most normal thing in the world? And what is the result of learning to be a passive watcher? By learning to watch, you are also taught deeply and subconsciously that you have no control over what happens around you. That is what watching after all is. You aren't supposed to participate.

You can watch a basketball game on TV and scream and cheer for your side, but no one (no sane person), thinks that they can in any way affect what happens on the court. You are destined to be a passive observer in this world, and this is the way it is supposed to be. This is a very very good way to control some 320 million people.

People who watch all the time understand innately that they are incapable of changing the things that they spend their lives watching. No one who watches a reality TV show or a murder mystery has the idea that if only he or she did something there in the living room, the young victim might be saved. Watching makes you the ultimate fatalist. You may see a breaking news story on TV or online about yet another school shooting. We are all outraged. But of course, no one really ever does anything. People may march for a day or two, but even as you watch them or even join in, you know in your heart that nothing is really going to change. You may hear pontifications from politicians and from the president but you also know that nothing is going to change. And, in fact, when nothing does change, you are not surprised. This, after all, is how the world really is. You even know, deep down, that you in fact have no control over events in your own life.

When seventeen students were shot dead at Marjorie Stoneman High School in Parkland, Florida, the media responded as it always does, with live breaking news. In the aftermath of the tragedy, the same rituals of speeches, pontifications, marches, and rallies were held—all on TV, all media events.

And when it was all over, did anything change? Of course not. Did anyone really believe that anything was going to change? Of course not. If you can't change the outcome of a Lakers game, what in the world makes you think that you can change the outcome of, say, global warming, or a war in Syria, or an election? You can't. So just sit quietly or cheer along (the way you do for the Lakers game). In fact, post your cheering on Instagram or Facebook. Hope that others give it a "like." Upset about some event in the world? Really want to effect a change? Give it a hashtag—#feelbetter?

EVERYTHING WILL BE FINE IN THE END

Passivity is bad enough, but two generations of watching taught us something much more destructive, and this is that in the end, everything will be fine.

Almost every movie you have ever seen, every TV show, every binge watched box set serial has one common factor attached to it. In the end, everything works out fine. The hero may be in trouble, the murderer may seem to be getting away with it, the fire in the high-rise may seem to be raging out of control, the hero and heroine may seem to be doomed, but, remarkably, in the eleventh hour and fifty-ninth minute, all is resolved—almost every time.

Watch this kind of amazing resolution of even the most seemingly intractable problem—an alien invasion, an asteroid headed for the earth, a serial killer on the loose—and you begin to be educated, over and over, that in the end everything will be fine.

This certainly gives us a sense of peace. This clearly is how the world actually works. And, you will note, that as passive observers, you did not have to do a thing, you did not have to lift a finger, save to refill the Doritos bowl, and magically, the world was saved.

Seventy percent of people now believe that climate change is real and represents a real threat to humanity. It could and most likely is the greatest threat that human beings have ever faced. Vast swaths of the planet may be rendered uninhabitable in the not so distant future unless real action is taken. Yet, people seem immobilized. This is because, I think, we have been educating ourselves for two generations now that in the end, every potential disaster turns out just fine without having to do anything about it. So when it comes to climate change, a disaster right out of a movie script, we deeply and instinctively believe that something will surely come along to solve this problem—probably right before the looming disaster strikes.

I did not come to these conclusions easily. I have spent the past thirty years or more in the media business. In the 1990s, I developed and produced one of the very first reality TV shows in America. It was a big success. It ran for more than ten years, and over time we fine-tuned it. What we discovered was the more shocking the content, the bigger the ratings.

We were not alone in this discovery. Radio shock jock Howard Stern discovered this for talk radio, as did Rush Limbaugh and many others. This is why The Learning Channel now runs programs like *My 600-lb Life* or *Dr. Pimple Popper*. It's why the Discovery Channel, which used to be about serious documentaries, now runs *Naked and Afraid*, why the History Channel runs *Pawn Stars*, why American politics and public discourse increasingly sound like shock TV. There are consequences to a media-driven society.

3

Mea Culpa

"Television is not the truth. Television is a goddamned amusement park."

—*Howard Beale in Paddy Chayefsky's* Network

I have spent most of my life in the television business. Even if you don't know my name, you may have seen some of my work. For many years, I made a great deal of money producing TV shows that exploited other people's suffering and problems and turned them into entertainment. I invented one of the first reality TV shows in the world, called *Trauma: Life in the ER*, and it aired for nearly ten years on TLC (the Learning Channel). What viewers were learning, I have no idea. But it rated. To this day, I still see it on cable channels up in the 300s, where no one ever goes.

To produce the shows, I put young people with small video cameras into hospital emergency rooms and filmed people who had just been in automobile accidents, who came in with knives in their heads or bullets in their brains, whose limbs had been torn off, and whose children had just died. We got them all.

When the ambulances pulled into the hospitals, the doctors ran to meet them, and so did we. Often, the victims of these horrific accidents and violent acts screamed and cried in both physical and emotional pain. We recorded it all. The network paid me $250,000 per show; I made more than a hundred of them. The networks, and apparently the viewing public, could not get enough.

The success of that series led to a dozen other series that were similar in format and content: police stories, fireman stories, paramedic stories. All people suffering or being arrested, their lives destroyed—reality TV at its finest.

I was very good at it. I had been in the news business before that, so I already knew what people liked: fires, shootings, wars, killing, bombings, danger, death, and destruction. I grew up in front of a TV set. My whole life revolved around screens—watching them and making the stuff that other people watched on them. Later, when the Internet came along, we just moved our stuff from TV to online. It was easy. I was like the Walter White of television—I knew how to manufacture stuff that would give a great high and be as addictive as hell.

I went to Manhattan parties with other people in the same line of work. We all laughed about what we did. It was great fun. And even though we were doing "reality," we would often see our work appear in the world of fiction.

A girl once came into an emergency room where we were filming with a live cockroach that had crawled into her ear. It took some time for the doctor on call to figure out how to extract it. About six weeks later, the same exact story line turned on up *ER*, the very popular fictional TV series about life in the ER. That version, starring George Clooney, was a big hit.

We were all having a great time, except of course, for the people whose arms had been ripped off or whose children had been maimed in a fire. But that was not our problem. It had not always been my plan to make money on the suffering of others and to package and sell it as entertainment. I had set out to be a journalist. Life just took a different turn.

I had gone to the Graduate School of Journalism at Columbia University, and after I graduated, I worked for the local PBS station and, after that, for CBS News, the network. I climbed rapidly in the news business, and in only a few years, I was a producer on *Sunday Morning*, one of the two CBS network news flagship shows, the other one being *60 Minutes*. I had arrived fast. And because I had had such a stellar career, I had returned to Columbia University as an adjunct professor to teach there as well. They liked having people who were working in the business, particularly recent grads, to teach their courses. They didn't have to pay us much. We had other jobs.

While I was teaching there, I met my future wife (and future ex-wife), who was a graduate student, but not in any of my classes. Let's call her Susan.[8] She had spent years working overseas for a major wire service. Going to Columbia was a career move for her, and she said she wanted to be a documentary filmmaker.

So early on in our relationship, we set out to make a documentary film. It was going to be about hospital emergency rooms. I had, in fact, heard a radio piece on NPR by Scott Simon, the host of *Weekend Edition*, on the same subject. I thought it would make a good documentary. I had studied documentary filmmaking when I had gone to Columbia myself, and if Susan wanted to be a documentary filmmaker, who was I not to assist?

And so it was that we found ourselves sitting in the emergency room of the University of Pennsylvania Hospital one Saturday morning, clutching our video cameras and hoping to make a documentary film, or at least trying to. Now, the essence of any good documentary, in fact the essence of any good film or video or TV show, is not so much the subject matter. This is a classic beginner's mistake. The essence of any good film is the characters. If you can make the audience care about the characters, to get involved with them emotionally, you can tell a great story. Every great film or video or TV series is about telling a great story, and every great story is about a great character. This has been true since the time of Homer, and it is just as true today.

As we sat in the waiting room at the HUP (Hospital of the University of Pennsylvania), we were looking for our characters, the stars of our documentary film.

And sitting across from me, as it happened, was our first star. He was a young man, probably about twenty or so, big and tough and

---

8 Names have been changed.

from the street. He looked scary. In fact, he was scary. But scary people also often make good TV.

The secret to great casting is the ability to talk to anyone, and more important, to make people feel comfortable so that they want to talk to you. Everyone has a story to tell. Your job is to help them tell their story. Often, all you have to do is ask, which is exactly what I did.

"What are you doing here?" I asked. It seemed a reasonable question, as we were both sitting in the ER. He looked at me for a minute, sizing me up. But I smiled and waited.

Then he leaned in close to me, as though to deliver a secret, and in a low voice he said, "I was shot six times."

Well, this is not the kind of thing you hear every day, but I did not break eye contact with him. Instead, I simply said, "You were?"

"Yeah," he replied. "Wanna see?"

I said that I did, and he proceeded to lift up his shirt. I supposed I was expecting to see gaping bullet wounds with blood pouring out, but instead, he showed me six grey lumps on his stomach and side. He had been lucky, considering the circumstances. He had been shot with a small-caliber gun, and the bullets, entering hot, had lodged themselves in his flesh and now, after a few weeks, were working their way out—like a splinter.

He asked me what I was doing there, and I said that we were making a documentary film. We showed him our cameras. We asked if he would be interested in being in our documentary. He looked over at his girlfriend. She nodded approval, and he said he would.

Just then, the triage nurse on duty called him and asked him to follower her into one of the bays on the emergency-room floor. His girlfriend accompanied him, and we asked him if it would be OK if we came along and filmed all of this. He said he was fine with that.

The four of us went into the ER, and as he lay down on one of the examining tables in a bay, the curtain pulled back. A young resident in a white coat joined us. She proceeded to examine him. We explained that we were making a documentary film and asked if it would be OK if we filmed whatever happened. She said, "Sure." She then took a paperclip and tapped one of the grey bumps.

"Do you hear that?" she asked me. It was indeed the sound of metal on metal. "I am now going to remove the old bullets." People love to explain what they are doing when you are filming. It makes life so simple. Everyone you film has already seen tons of TV and movies. They all know exactly what you are looking for, and more important, what they are supposed to do. We are the best TV performers in the history of the world. And with that, she picked up the forceps or something (I am no doctor) and proceeded to start tearing out the bullet with no warning and no anesthetic.

Needless to say, our newfound friend began to scream in pain. It was a wail of pain the likes of which I had never heard. Something so base, so intense, so very real: "Gaaahhh!" And, much to my delight, we had caught it on camera.

The young resident doctor paused for a minute, looked him in the eye and said, "You big baby." We kept the tape rolling. She was going to make great TV. And so was he, if he survived.

Patients like this were called frequent flyers in the ER. They would get shot up. The doctors would fix them up and send them out on the streets, where they would get shot up again. She was a bit fed up with him, which is why I think she didn't administer any anesthetic. It was a kind of rough justice she was administering all on her own.

"There," she said, "doesn't that feel better?"

"That do *not* feel better," said our patient. He also, rather kindly, made eye contact not with the doctor, but with the lens of my camera. He too was already performing for his prospective viewers.

The doctor held the spent bullet, trapped in the forceps before our patient to show him what she had removed. Then, suddenly, and very much without warning, his girlfriend reached across the table upon which he was lying, snatched the bullet from the forceps, and held it before her eyes, much as a jeweler examining a fine diamond. "You said you was shot with a .38," she pronounced. "This ain't no .38. This is a Glock 9 millimeter." You could not have scripted it better.

I glanced over at Susan. We were recording? She nodded. This, in the media business, is what we call documentary gold. It's these kinds of little nuggets of reality that just jump off the screen and that people remember.

We spent the weekend at the hospital, shooting whatever came in the door. One of the advantages of filming in the ER is that I really did not have to do a lot. The cases—a knife in the head; an automobile accident victim; head clipped by bus mirror; man who fell down in a bathtub, broke off the ceramic soap dish, and sliced a two-foot gash in his back—all just came in through the door, and all we had to do was point the camera and hit Record.

By Sunday night we had accumulated a massive amount of footage, and now all we had to do was edit it together. Here, we encountered our first problem.

We went home and started to screen through the footage we had shot. The next step in documentary filmmaking is to log your material, then write a script. Having cut my teeth at both PBS and CBS News, I had a pretty good idea of what the script should be. I could almost hear it in my head: *The battlefield conditions of Vietnam came home to the American city of Philadelphia last week, and so did the medical lessons learnt on that battlefield.* We'd have lots of dramatic music and narration by Will Lyman (the voice of so many *Frontline* Emmy Award–winning documentaries on PBS). When I laid this all out to Susan, she laughed in my face.

"That sucks!" she said. "I'm not doing that."

This struck me hard. I was a professor and professional video journalist, and she was a beginner. But it was early in our relationship, so I was not going to get defensive. Good television producing is all about pretending to listen to the rest of the team and then doing what you damned well want to do anyway. So I said, "Well, what did you have in mind?"

She did not hesitate. She had a very clear vision. "I want to do a lot of fast cuts with some really cool rock music."

Here I took a deep breath. I asked myself, *What is more important here: the made-up, totally fictional "documentary film" we are supposed to be making, or moving the relationship along?* I would certainly have opted for the moving-the-relationship-along choice, no question, except for one niggling problem. Frankly, I did not have much hope in ever selling our "documentary film." Documentaries are notoriously hard to sell.

Now, you will remember that we spent the weekend in Philadelphia, even though we lived in New York. As it turned out, the former head of programming for WHYY, the PBS station in Philly, a guy named John Ford, had just been appointed the head of programming for the Learning Channel. The network had just been bought by Discovery, which was itself not all that old or established. The Learning Channel now calls itself TLC, much the same way that Kentucky Fried Chicken now calls itself KFC, to mask both the Kentucky and the Fried part (both remarkably unhealthy). For someone who wants to create a profitable TV franchise, the last thing you want to call it is "Learning." Does learning sound exciting to you? Well, exactly.

In any event, I had wrangled an invitation from John Ford to send over a bit of our documentary if I was so inclined. Now, in my mind, here was John Ford, who had come from PBS and was now running something called the Learning Channel, and he wanted to take a look at a bit of our documentary about a hospital in Philadelphia, his former hometown. Well, of course he was looking to bring that kind of PBS quality to his new job at the Learning Channel. So the last thing I wanted to send him was some student film. Rock music cut against people with knives in their heads or bullets in their chest—was she out of her mind?

So I asked, "Are you out of your mind?"

This did not, as you might imagine, go down particularly well. Her response was cogent and to the point: "F**k You! Your ideas are old and tired!"

So I responded in kind, carefully weighing out that balance between nurturing the relationship and selling my very first documentary film. So I opted for the more likely prospect—the relationship.

"OK," I said, throwing in the towel. "Go ahead. F**k it up! Learn the hard way."

So she did. Well, at least she cut it exactly the way she wanted to cut it. It no longer bore any relationship to any kind of Emmy Award-winning documentary film I had ever seen. Instead, it was closer to the kind of music videos that MTV was just starting to run. Lots of driving guitar and bass. Pounding, loud, then a piercing scream— "My baby is going to die!"

Back to the music rhythm line, another scream—"Don't cut off my leg!!"

Then, just a piercing scream—I think it was the guy who had the bullet pulled out of him.

When she was done, she showed it to me. It only ran about five minutes.

"Well," she said, "what do you think?"

I took a long pause, then said, "They are going to throw us out of the room."

And so it was, with a great deal of trepidation, that we got on the Amtrak train down to DC to take our meeting with John Ford and the team from the Learning Channel. I thought the meeting would be over very quickly.

We arrived at the offices of the Learning Channel, and, at 2 p.m., the management team for the new channel all filed into the conference room to see what we had to show.

"Well, let's see what you've got," John Ford said, after the requisite round of shaking hands and introductions. So we slid the tape into the VHS deck, watched as the machine swallowed it and cued up the tape, and hit play.

I already knew what was coming, so I had packed up my stuff and prepared to beat a hasty retreat to the door. The room went dark, and the video, which I had already seen more than enough times, began to play.

"My baby is going to die!"

I looked around the room. Cast in the blue glow of the projector, the TLC team looked even more ghastly than they did in fluorescent. You could see their jaws dropping as the video began to play.

"Don't cut off my leg!"

I could hear Ford already screaming at me, *Are you out of your mind?*

The five minutes seemed to drag on forever. Then, finally, it was over.

I looked at Ford. He stared at me. Then he said, "I would like to order thirteen half hours of this series, at $200,000 per half hour. Do you think you can do that?"

Dumbstruck, I nodded in assent as $2.6 million slid across the table to us. I didn't even know this was a series. It was supposed to be a documentary film. But now it was a series, or at least it was going to become one.

It was going to be called *Trauma: Life and Death in the ER*, except that they thought that calling it *Death* might be a bit negative. So it was renamed *Trauma: Life in the ER* instead. As we walked out of their offices and hit the street in Bethesda, I turned to Susan. "Well, you certainly knew what you were doing," I said.

She looked at me for a moment, than answered, "Yes. And you don't. So stay the f**k out of my edit room."

About ten years later, my divorce lawyer would say to me, "When did you start to realize that the marriage was not working out?" In retrospect, I would say, it was probably at that moment.

In any event, *Trauma: Life in the ER* went on to become the kind of smash success that TV producers dream about. The fact that it was the very first show that we did was nothing short of amazing. The network could not order or run enough of them. And that was only the beginning. Because we had a hit show our first time out of the box, John Ford kept coming back for more ideas for more series. Eventually, we had about eight major series in production simultaneously: trauma, paramedics, police story, labor and delivery. Within eighteen months, I was running the largest nonfiction television production company on the East Coast.

But what was it that the network wanted? It wanted what people wanted to see. And apparently what the people wanted to see was blood, suffering, pain, screaming, crying. And we were not alone. The more I looked around, the more I could see that it was not just TLC; it was every network. They all wanted the same thing. The idea was to grab people by the eyeballs and hold their attention so you could sell commercials to the advertisers. That was it.

The *Trauma* series was doing very well, and we were shooting *Trauma* shows all over the country. Wherever there was a hospital with an emergency room, we sent a crew. The series was the network's number one show, year after year. People could not get enough of it.

Around the fifth or sixth year, we shot a couple of shows in San

Diego at a hospital there. Our camera crews would pretty much live in the hospital. We even kitted them out in scrubs so that they could blend in—also so that they could slide into the operating room in those really great situations where the ambulance pulled in and the poor bastard who had gone head first through a car windshield went directly to surgery.

One day, while shooting in San Diego, an ambulance pulled up, lights flashing. This was the signal to start filming. You never knew what was coming through the door. What came through the door on this day was a seventeen-year-old kid who had been car surfing. Car surfing is when you stand on the roof of the car while your friend drives it around. (No one ever said that seventeen-year-olds are not idiots.)

In any event, this kid had fallen off the car and hit his head on the cement. When the ambulance arrived, he was out cold, and when his parents arrived, we cornered them with the cameras. But first, we asked them to sign our standard release form. No one gets on air without signing a release. Then we filmed.

In the beginning, it looked like the kid was going to pull through—just a scary story, not much from our point of view. Over the next few days though, the situation for the kid turned worse. Instead of waking up, he slipped into a coma. As he was a local, there was a long line of visitors—grandma, sister, football coach, the mother's eternal vigil by the bedside. We shot it all. It was getting to be better TV all the time.

Then he was brain dead and on life support. Now, we had some action (so to speak). The parents had to pull the plug. Of course we filmed that too. Is was a very emotional moment, and for the parents, devastating. For us, it was just another day at work.

In any event, we went back to New York and the show together. It was only one of our thirteen shows for that season. But the car-surfer-kid story was so good that we made it the spine of the show. That's when we start the show with the kid being brought in and then keep coming back to him as the hour progresses. Then, in the end, when he died, that's what we put at the end of the hour. It's called bookending.

We cut the show and delivered it to the network and they *loved* it. It had everything that makes a big ratings success—love, families, disaster, crying, and then the end: perfect

They loved this particular show so much that they decide to make it the season opener. That's a big deal; it gets a big ad campaign. They wanted to go all out, so they decided to have a big season launch blow-out party in San Diego. They had an ambulance and rescue helicopter arrive. The local press, of course, was tipped off or paid off to publish stories about the "exciting new season" and the great San Diego show that was going to kick it off.

Susan and I were actually at home in New York when the first call came in. It was from a Mr. Christianson in San Diego.

"Do you know a guy named Christianson in San Diego?" I asked Susan.

She shrugged her shoulders but took the call.

A half hour later she came into the room, ashen-faced and looking devastated. Something was clearly up.

"Do you remember," she said, "that kid who was car surfing in San Diego?"

It took me a quick moment. "Oh yeah, show 607."

"Right," she said. "Well, that was the dad. Do you remember him?"

"Vaguely," I replied. I had seen so many parents watch their children die in the ER in the past few years, it all became a blur.

"Well, he said that he read about the season premiere in the local paper in San Diego, and he wants to stop the show."

"What do you mean, he wants to stop the show?"

"He doesn't want it to air. He said that he doesn't want his son's death used for public entertainment."

"What did you tell him?" I asked.

"I told him that he had to call the network, that they owned the show, not us."

"Well, that was the right answer."

Within a few hours, Mr. Christenson called back. He had called the network. They didn't care. They said he had signed the release. They said, effectively, "too bad."

For the next few days, Mr. Christenson called us four or five times a day, every day.

"What kind of creatures are you people?" he asked.

He had a point. We were complicit in something terrible. We were using other people's suffering for amusement, to make a profit. It was an electronic version of the Middle Ages, dragging out the physically deformed for public entertainment. And slowly, over time, I came to realize that this kind of commercial exploitation, this warping of reality to turn everything into entertainment that "rated," was not limited to my little series. It was the way the entire media industry worked.

People watching TV shows or playing video games or on Facebook or Instagram or YouTube or anywhere else for that matter, may think that these programs have been created with the interests of the viewer in mind. But that would be wrong, dead wrong. The viewer or the user of the app is not the client. The client is the advertiser. The viewer or the user is the product that the media or tech company is selling to the advertiser.

The media business and the tech business are not there to serve you. They are there to serve the advertising world. You, your attention, your time, your eyeballs, are what the media and tech companies are selling. And the media companies and the tech companies will do anything they can to attract as many eyeballs as possible, no matter what the consequences. They have no interest in your education, your well-being, your enrichment. They are the Colombian drug cartel of the digital age.

So what happens to a society that spends four or eight or even eleven hours a day, every day, watching this kind of stuff? That is, murders, kidnappings, violence, crime, shootings, and car crashes, all real or fictional, over and over and over for years on end? In the next chapters, we are going to look at the often-unintended consequences of technologies. As we move into the twenty-first century, we are going to be faced with other new and very disruptive technologies from artificial intelligence to gene manipulation and more. Perhaps it would behoove us to take the time to examine the potential impacts of new technologies before we plunge headlong into them.

4

Electric Heroin

"The 'well-informed citizenry' is in danger of becoming the 'well-amused audience.'"

—*Al Gore*

In 2001, done with making reality TV, I opened a video bar and café on the Bowery, in New York's lower east side. It was right across the street from CBGB, the music club that had given birth to the punk-music revolution. You could not ask for a better location for a new venture. I wanted to be hip and edgy.

Today, the Bowery has been gentrified. It has expensive glass condominiums, fancy restaurants, and a Whole Foods. When I was there, there were SROs (single-residency-occupancy hotels), homeless people, and food banks. But real estate was cheap, and it was near NYU, where, having left Columbia, I was now teaching.

I wanted to open a bar and café where we would give out video cameras, have edits, and show videos and movies that people who were members would make and share. It was the antithesis of what I had been doing. Think of it as a physical kind of YouTube long before there was a YouTube, and this was way before any kind of social media ever existed. It was almost before the Internet. As a money-making business, it turned out to be a disaster. But it led to something far more interesting.

One day, a very big man named Jamie Daves wandered into my place. He wanted to talk about the future of television. As I had coffee, muffins, tables, chairs, and time in profusion (business was slow, and the muffins were not moving so well either), I pulled up a chair and let the coffee and muffins flow. We must have chatted for an hour or so about the video revolution and all I thought was going to happen in the world of video. Then he left.

I did not think of Jamie Daves again until a few days later when he called. "I represent former Vice President Al Gore," he said, "and the vice president would like to meet with you."

This was early in 2001, and Al Gore had just lost the election for president of the United States, so the idea that Al Gore would want to meet with me was nothing short of astonishing. Of course, I said yes.

"The vice president wants to know if you have a private place where the two of you could meet," he said. I told him I lived in a loft in Soho and we could meet there.

So I went home that night and told Susan that Al Gore, former Vice President Al Gore, was going to be dropping by tomorrow. She looked at me with a look of utter contempt, rolled her eyes, said, "Yeah, right!" and walked away.

The next morning, at eight, my doorbell rang. In a Soho loft people don't get a doorman—at least they didn't in those days—but we did have a video camera so that you could see who was at the door and buzz them in. And there, on my little black-and-white screen in the kitchen was, in fact, former Vice President Al Gore. It was just like watching TV.

"Al Gore here," he said, in a voice that I recognized immediately. A few minutes later, he was at my door, yellow legal pad in hand, pens in a pocket protector in his white shirt. We shook hands, and I invited him to sit at my dining room table.

We chatted for a while. In all honesty, I was a bit gobsmacked to have the former vice president of the United States, a man I had voted for for president, sitting at my table. I thought about saying how I thought he had been ripped off in the election, what with the hanging chads and the Supreme Court and all, but I let it slide. Instead, I asked him what I could do for him. He took a long pause, then, referring to his yellow legal pad said, "I'm thinking of starting a TV channel. It's going to be about history or politics."

"Al," I said, "you're making a big mistake. There's a revolution going on. People are starting to make their own stuff in video."

I didn't really know whether there was actually a revolution going on in video or not, but that idea had been the basis for the video bar and café, so I figured, why not? Gore took assiduous notes as I talked. I told him about Gutenberg, and how the "revolution in video" now was like Gutenberg's democratization of the world of print five hundred years before. He kept writing. Almost immediately, he got it. I asked him, of all the things he could have done, why he wanted to get into the TV business, and it was one of those explanations (as when he would later explain global warming to me) that changed the way I saw the world and how I thought.

I knew that television was an incredibly popular and powerful medium. I had worked in it for years already. What I did not understand, at least until then, was *why* it was so popular and so powerful. It was addictive, he said. It wasn't about the content so much as the physicality of the medium itself. Television and video and film, he said, were really no more than flashes of light on a screen. But the very fact that they were flashes of light on a screen tapped into something very deep and very powerful in our DNA.

AL GORE EXPLAINS IT ALL TO ME

For the millions of years that we were evolving as a species, and later as we stood erect for the first time in the savannas of Africa, a flash of light in the tall grass or the forest meant one of two things—either there was something to eat out there, or something that was going to eat you. You did not get a second chance in nature, but those flashes of light were a kind of warning. Eat or be eaten; run for it or run away from it. But in either case, it meant, *pay attention—this is very important.*

Competition both for food and for survival was intense in those days and really was a matter of life or death. And, as it turned out, only those early humans who had the most finely attuned sense of responding quickly to those flashes of light either ate or survived long enough to reproduce.

So slowly but surely, over generation after generation of natural selection, we evolved to have an extremely fine-tuned sensitivity to flashes of light. We could not help but look toward them. They commanded our attention. Even later in human development, when we were not so much on the hunt for animals or running from them, but still battling with swords in hand, again this exquisite sensitivity to flashes of light was often the indicator of whether we would survive to fight another day and reproduce or be eliminated from the gene pool. Those who responded the best and the quickest were the survivors and our ancestors and progenitors.

That extremely important and deeply held response to flashes of light—*Look! Pay attention! This is important!*—for the most part lay dormant in most of us, but it was always there, like the shingles virus, lying in wait. Go to Las Vegas sometime and walk into any casino and what do you see? Flashing lights. They are there for only one purpose: to get your attention. An ambulance or a police cruiser or a fire truck goes tearing down the street and wants to get your attention—flashing lights. Your car is in trouble on the side of the road—turn on the flashing lights. You can find hundreds of examples, once you start to look for them.

Now take a look at the screen on your smartphone. When it wants to get your attention, what does it do? It starts to flash a small icon. You have to look. You cannot resist. Before the interest in global warming, Al Gore understood what made television and video and on-screen media so irresistible. It was addictive. It was the flashing lights.

A few years ago, journalist Paul Miller wrote a piece for the Verge entitled "I'm Still Here: Back Online After a Year Without the Internet."[9] Miller had taken it upon himself to boldly go where no one, apparently, wants to go anymore—offline. For most of human history, of course, we all got along quite well without the Internet, a

---

9 Paul Miller, "I'm Still Here: Back Online after a Year without the Internet," Verge website, May 1, 2013, https://www.theverge.com/2013/5/1/4279674/im-still-here-back-online-after-a-year-without-the-internet.

device that has only been around for the past twenty years or so. But the fact the the Verge would even commission such a piece, not to mention the enormous attention he got when he emerged, demonstrated the astonishing degree to which we are all addicted to the media in its various permutations.

His story was covered everywhere—ABC, CNN, the *Atlantic*, you name it. You would think that he had walked across Antarctica unaided in the dead of winter or had climbed Mount Everest. What made Miller's year-long expedition into the unknown so newsworthy was that it is now almost inconceivable that an individual would be able to give up his or her addiction to something that did not even exist twenty years ago. If someone gave up cigarettes for a year, do you think that person would write a book about it? Do you think that ABC News or CNN or the *Atlantic* would call it news? As it turns out, the media is more addictive than nicotine—a lot more addictive. We are absolutely gobsmacked when someone can give up social media for twelve months.

Here's one I like even better. It was on the *Today Show*. Teacher Dave Heywood at Black Hills High School in Washington State asked his students to give up their phones for a week. A week! And it makes national news! And what were the results? "I started reading a book last night," says Jessica, one of the students.

No part of the media is more addictive than the flashing lights and sounds of video in its various permutations, from TV shows to video games. Anyone who has ever been to an AA meeting or been in rehab can tell you that the very first step to curing any addiction is simply being able to admit that he or she is in fact addicted. Denial is the hallmark of the addict. It is often easier for a friend or relative to recognize a problem drinker than it is for the problem drinker him- or herself. One glass of wine a day is fine, or even two, but once you start to get up there to five or six a day, not to mention the scotch or the vodka, well, then you can begin to think there might be a problem.

So, going to the movies once or twice a week was fine, but spending eight hours a day, every day, in a movie theater might be seen as a kind of a problem. First it was black-and-white TV sets in our living room and programming that actually ended at midnight. Now, it's a seventy-two-inch massive screen on every wall in every room in the house. And, if that was not enough, we carry a smaller screen with us wherever we go, and we get very very anxious if we are separated from it. This is a problem. This is an addiction.

You can like sugar, and our desire for sugar is, like our desire for stories, deep in our DNA. If you found a cache of honey while you were out hunting twenty thousand years ago, and you downed it all because its really concentrated energy and would serve you well in the lean days ahead, that's great. But if you start downing eight boxes of Honeycomb cereal, donuts, and Ring Dings all day long, every day, you might have a problem.

We are what addiction specialists call "high-functioning addicts." This is not falling-down-in-the-street drunk, but rather the alcoholic who is able, at least in his or her own mind, to manage the addiction. One of the telltale signs of a high-functioning addict is that the person tends to hide his or her addiction from others—the vodka bottles hidden in the desk drawer or under the car seat. Alcoholics tend to hide their bottles because they are ashamed of their behavior. They know it is wrong, but they just don't want to be confronted with it. When it comes to our own addiction, we are no different.

Even though the average person spends four to eight hours a day watching TV or videos, even though watching video in general and TV in particular is now our number one activity, how often do you see people on TV shows or in the movies watching TV? You'll notice that there is no TV set in the *Big Brother* house, for example. That is because if there was one, all of the cast members would sit all day, watching TV. That, after all, is how most people pass their time. But no one would find it particularly exciting to watch other people watch TV. Being confronted with the stark reality of how we are actually spending our lives would prove too depressing, so like the alcoholic hiding the bottle, we pretend it is not happening—not to us.

The people in the fictional world of reality TV or dramas or movies in which we lose ourselves on a regular basis have *interesting* lives. That, of course, is the fiction. Unlike the rest of us, they actually *do* things. Which then is the real reality?

And if you think this is bad for you, the future bodes much worse.

Your children are going to be spending almost all of their lives not just watching but immersed in fake stories. With the advent of the next generation of watching technology—AR (artificial reality) and VR (virtual reality)—we are going to be entering into an entirely new and uncharted world of immersive content. If you think 2D video is addictive, wait until you try the hypnotic charms of immersive reality. This is coming.

This kind of addiction drains you spiritually. It desensitizes you to the pleasures of real life. Over time, it numbs your sensibilities. Because you are continually exposed to increasing levels of intensity in your observed life, you become increasingly unable to feel things in your real life. Bathe in a swimming pool of cheap cologne every day for a year and you will soon lose the ability so smell or appreciate a single rose that you might come across. It deadens your senses. Bathe in an ocean of never-ending amped up stimuli—wars, bombs, car chases, shootings, mass shootings, terrorism, fires, disasters, murders, serial killers—things that are actually completely divorced from the true day-to-day realities or your life, and how soon before you cannot feel a quiet walk in the woods, hear a bird song, or taste an apple? You become overloaded, calloused to your own reality, which suddenly seems so small and inadequate. You are inured to the actual small pleasures in life because you can no longer really feel them. Fed by a never-ending stream of ever greater stimulation, the actual real world seems increasingly insufficient. Hence you and your real life are failures.

A few years ago, my wife Lisa and I were at an art show at the Armory in New York. There were literally hundreds of galleries showing thousands of pieces—paintings, photographs, prints, sculptures. There were only a few video installations, yet each of them drew a massive crowd. No matter how bad or mediocre the video installation was, it drew, by far, the most massive crowds—far more than the paintings or the sculptures or the photographs. Why was that? What was it about the medium itself that was like a magnet?

I was walking across Times Square recently, as fast as I could, when I was suddenly caught up in a crowd that was not walking anywhere. This is a perpetual problem for New Yorkers—tourists who just stop in the middle of the street to gawk or take a selfie or something. But this was not just one tourist or two, this was a group of several hundred. And they were all staring up at a giant screen on the side of a building. And what was on the screen? It was video of . . . themselves.

They were mesmerized staring at themselves on a giant video screen. They waved. The images on the screen waved back. They took videos of themselves in a video. They could not look away.

It was, in a sense, the perfect metacommentary on our drunk-on-media society. We are addicted to watching videos, and our favorite thing to watch, as it turns out, is ourselves. It's not our real selves—that would be too depressing—but some kind of video-mediated representation of ourselves as we would like ourselves to be. All of our programming is about perfect people living perfect lives, so why would we not emulate the endless content that we now dub "reality"?

The people in Times Square were all there physically. If you wanted to see them, all you had to do was open your eyes. But no one was taking videos of the crowd on the ground. Put those same people on a fifty-foot screen as a video representation of exactly the same reality that was happening at that very moment on the ground, and immediately you had a video simulacrum of the selfsame event. On the ground it was banal. On the screen, it was magical.

You can see this phenomenon on YouTube or Facebook or Instagram continually. You may go on a vacation, but the vacation is not real until it, in its perfection, is captured in video and then reproduced on other people's screens. We are all starting to live in a mediated reality.

Al Gore, as it turned out, was prescient not just about climate change but also about the very addictive nature of video. Long before YouTube or Instagram or Facebook or Netflix, he saw what was coming.

THE RISE OF THE SELFIE CULTURE

Today, there are 1.9 billion people watching videos on YouTube: one out of every four people on the planet. Five billion videos are watched on YouTube every day. YouTube is also the second largest search engine in the world, after Google. What makes this even more extraordinary is that this video-watching phenomenon is not limited to YouTube. YouTube may have been the beginning, but now almost every social media app is rapidly incorporating video. This year, Facebook will in fact surpass YouTube in volume of video. Facebook gets an astonishing eight billion video views *per day,* and more than 100 million hours of video are viewed on Facebook *daily.* Instagram, Twitter, Snapchat, and every other platform flood the Internet with billions of videos a day. If video is indeed addictive, and it appears to be, we are a nation awash in visual heroin.

The rise of the selfie is a manifestation of our addiction to images and flashing lights, but also speaks volumes about the impact of screen culture and screen addiction on society.

The selfie was a virtually unknown phenomenon until Steve Jobs introduced the backward facing lens in the iPhone 4, allowing you take a picture, and later, a video of yourself, without all the bother of a tripod and a cable that prior conventional photography and video once required. Once a selfie was possible, the genre took off with a vengeance.

Today, according to the Pew Research Center, more than 91 percent of teenagers post selfies with regularity. Initially in still photos and now in video, the self obsession of the content dominates sites such as Facebook, Instagram, and Snapchat. Why is that?

Dr. Aaron Balick, a psychotherapist who has written extensively on the subject, theorizes that the Internet allows us to create what he calls "active online identities" and "passive online identities." The passive identity is the one that people find when they search for you, often in postings from other people. It is an identity that you have no control over. The active identity, on the other hand, is one that you create entirely for yourself, and it can—and increasingly does—have little resemblance to reality. It is the "self" that you wish to project to the rest of the world.

This is an exploding trend, and it bespeaks a consequence of a world awash in created false images. The endless exposure to other people's "active identities"—largely fictional—has a tendency to drive people, particularly the young, into a depressive state, feeling their own life is inherently inadequate. In response, they then create their own fictionalized version of reality.

In the end, no one lives in the real world or lives a real life. Everything increasingly becomes illusion. The selfie and the selfie culture it created is, in many ways, a harbinger of what is yet to come when VR and AR reach full force: entire lives lived in fictional and created realities, all made of flashing lights.

In his seminal book, *Four Arguments for The Elimination of Television*, Jerry Mander points out that 80 percent of the US population lives in cities. That is, our lives, for those of us in the 80 percent, are completely artificial and completely mediated. Our food is industrially processed; the air we breathe is industrially processed; our water is industrially processed. We are, in effect, completely alienated from the smells, sights, sounds, and experiences that formed us over a million years. And now, our experiences are no longer real either. We experience the world not through touch and taste and smell, but by watching it on a screen. Experiencing the world through events on a screen only creates experiences as false and as mediated as the rest of our hermetically sealed and artificial urban environments. And what is worse, we come to accept this as real.

Our idea of nature is now *Blue Planet* or *Nature* on PBS or BBC. We don't feel the natural world, we don't taste it, and we don't smell it. We just perceive it as a perfect image. The people we see in movies and TV shows are as perfect as the animals in *Blue Planet*. Tourists to Yellowstone try to pet wild bison because they seem so nice on TV. The world population of sharks is decimated because of *Shark Week*. The perfect people we see on *Real Housewives* or the *Bachelor* become our model of how all our lives should be lived and what we should look like. They are, increasingly, our only perception of how life should be lived, and thus we spend our lives emulating what is, in fact, pure fiction.

TOO MUCH INFORMATION

They say one picture is worth a thousand words, but according to Forrester Research, one minute of video is worth 1.8 million words. If we are watching five hours of TV or video a day, then we are consuming the equivalent of 540 million words a day, or 197 billion words a year. The entire *Encyclopedia Britannica* has 44 million words in it, so we are effectively consuming the equivalent of eleven *Encyclopedias Britannica* a day. That is an overwhelming amount of material. And consuming all that content has an impact on us too.

We are now continually awash in a veritable ocean of information and content. This too is an entirely new human experience. No human, in all the history of human beings on this planet, has ever been exposed to so much information, so many images, so many sounds all the time. If you think back to our medieval ancestors, they could and generally did spend their entire lives without ever seeing a single book or, for that matter, as much as a single printed page—let alone a photograph or the movie *Star Wars*! This was the human condition for all our history, since the ancient Greeks, since we were living in caves.

For our medieval ancestors lived in what would be for us a veritable visual desert. One can only imagine, for those who made the pilgrimage to a cathedral like Chartres Cathedral in France, what the experience must have been like. The walk could take weeks or even months, but it is today almost impossible for us to comprehend the astonishing power that the stained-glass windows at the Chartres Cathedral would have evoked once they had finally arrived. Seeing it—the color, the light, the images, the sheer overpowering size—was unlike anything in their prior experience, their world being devoid of any kind of visual images of almost any kind. One can readily understand how they must have dropped to their knees and crossed themselves and believed they had experienced a miracle. And in a way, they had—the miracle of visual stimulation.

The stained-glass windows at Chartres was the *Star Wars* of its day times one thousand. It wasn't that people were then starved for information; it was rather that they were all living the normal human experience—a world without mediated imagery. And make no mistake, mediated images are powerful. They are enhanced and sanitized reality.

There is a reason that in the Ten Commandments, the second commandment, right after Thou Shalt Have No Other Gods Before Me, is Thou Shalt Make No Graven Images—even before not killing, not committing adultery, and keeping the Sabbath holy. Although adherence to the ban has varied according to the school of Islam, the religion has historically avoided the use of images of sentient beings, or even all living beings. That is why the religion became fascinated with calligraphy. Such was the power, and the fear, that imagery possessed.

Now we live in a world of nothing but images. If we go to the Chartres Cathedral today (not really all that high on any tourist's list anymore), we might spend a few minutes there. We might look up at the famous stained-glass windows.

"Nice!" we might say to whoever is with us. "Really amazing." We might take a photo, or more likely, as selfie of ourselves, and post it on Instagram or Facebook. "The famous stained-glass windows at the Cathedral at Chartres" we would write, to let all our followers know we were in France and that we had a modicum of culture. "Amazing!" we would write. And that would be that, except for a few likes—on to the next tourist event.

In little more than one lifetime, this thing that no one had ever experienced before in all of human history, this activity of perpetually watching things, is suddenly the dominant influence in our lives. And yet, in what may be the strangest aspect of this, almost no one has really paid all that much attention to it. Like the alcoholic who says, "I can stop any time I want to" or the heroin addict saying that he or she is in control, we bury the truth of our addiction.

Go into any bookstore and find the television shelf. What's on the shelf, this medium that has seized our culture and our brains and our lives? *The History of* I Love Lucy, *The Best Soap Operas of the 1970s, The Big Book of Jeopardy.* Try to find some serious analysis of one of the fastest and greatest transformations in human history, and you will have a hard time finding anything. It's almost as though we don't want to know about it.

THAT'S ENTERTAINMENT

There is another thing that spending your life watching does. It pre-conditions you to expect and indeed demand to be entertained. If something is not entertaining, it is boring, and we are done with it.

As watchers, the only currency we have is in deciding what it is we will watch. If we don't like it, we won't watch it. The choice gives us some sense of control over our lives. It is a false sense, but it is there.

Thus, as no one is forcing us to watch anything specifically, we will naturally gravitate to that which is the most entertaining. And the makers of all that video content cravenly compete for our eyeballs; they will accede to our desires and continue to tinker, to experiment, to find the magic formula that keeps us coming back.

Thus, the media companies have provided us with an endless stream of entertainment. They spend billions on audience research to make sure we crave the entertainment that they are providing us with. They keep fine-tuning it, and when they find some format that works, they repeat it ad infinitum: reality TV shows, game shows, doctor shows, Marvel comic book heroes, endless Batman iterations, news shows, pornography, Facebook, Instagram.

As a result, there is no place for the boring, the tedious, the complex, or the difficult to digest. Everything must be amusing, entertaining, and resolve quickly. Real issues are complicated; they are the broccoli of content. No one wants to eat them. No media company in its right mind would even put them on the menu. Customers would run away screaming and never come back.

Entertainment is like heroin—you constantly need a bigger and bigger hit to get the same high.

PLATO GOT IT RIGHT THE FIRST TIME

If you went to college and studied Philosophy 101 as I did, then you will be familiar with the allegory of the cave from Plato's *Republic*. It begins as a dialogue, narrated by Plato, between Socrates and Glaucon, Plato's brother. Socrates asks Glaucon to imagine a cave in which men have been chained at the neck for their entire lives, so that the only thing they have ever been able to see is a blank cave wall. Behind them is a fire, and between them and the fire is a raised walkway upon which people move, carrying objects. The only thing the prisoners can see is the shadows of the objects that the firelight casts on the shadow walls. This is the only knowledge of the world that these prisoners have ever had.

Socrates then supposes that one of the prisoners escapes and is able to see the fire and the walkway for what it is. Socrates says that the man would be so freaked out that he would run back to the prison and the shadows—the world he knew and in which he felt most comfortable.

He then supposes that someone takes this man and drags him out of the cave and into the sunlight. At first, the light would be almost blinding, and it would take time for him to adjust, but slowly he would come to realize that the entire world that he had known was but shadows and illusion, that what he experienced now was the real world. Finally, when he was able to look into the sun itself, he would understand the harsh truth of the terrible world in which his fellow prisoners had been dwelling all their lives.

Socrates then supposes that this freed man would naturally want to return to the cave and free the rest of his fellow prisoners, to show them the light, so to speak. But returning to the cave, his eyes now adjusted to the sunlight, the escapee is no longer able to make out the shadows. His fellow prisoners conclude that his blindness is a function of his having left the cave, and all agree that they will kill anyone who tries to take them out of the cave in the future.

The day after I met with Al Gore in my loft in Soho, my cell phone rang. It was Gore. "What are you doing tomorrow?" he asked. Whatever I had planned, I was ready to cancel. He asked if I could fly out to Deer Valley Resort in Park City, Utah, and meet with his business partner Joel Hyatt. Hyatt's wife would also be there, along with Tipper. He wanted to know, if it was so easy to make videos, could I teach the four of them to shoot and edit and produce their own videos?

I was on the plane the next morning. I brought a few small video cameras and a few laptops for editing, and in a couple of days, everyone was out making videos. At the end of my stay, Al and Joel asked me if I would join the management team for the new TV channel they were going to make. I asked who else was on the management team. They looked at each other and said simply, "us."

I was in. Seven years later, we sold Current TV for $500 million. Al had been right. TV was a good business.

5

Running with Scissors

OR
THE UNINTENDED CONSEQUENCES OF TECHNOLOGY

"Progress has never been a bargain. You have to pay for it. Sometimes I think there's a man who sits behind a counter and says, 'All right, you can have a telephone, but you lose privacy and the charm of distance. Madam, you may vote but at a price. You lose the right to retreat behind the powder puff or your petticoat. Mister, you may conquer the air but the birds will lose their wonder and the clouds will smell of gasoline.'"

—*Henry Drummond, a character in* Inherit the Wind,
by Jerome Lawrence

When Prometheus stole fire from heaven and gave it to man, his creation, it angered the gods and Zeus. It angered Zeus because he knew that with the power of fire, man would be able to work metals, have science, and one day perhaps become even greater than the gods themselves. Zeus, as it turned out, was right. Technology has consequences.

As a species, we have a love of any new technology; we embrace it without considering the consequences. When the first automobile was invented by Karl Benz in 1885, we looked at it and said, "This is great! I have to have one of these." One automobile was fine, but today there are more than one billion automobiles in the world, each of them consuming oil and spewing out hydrocarbons and six billion tons of CO_2 every year. There is a price to pay for every new technology, whether we want to face that fact or not. Nothing comes free.

There may be a price to be paid for every new piece of technology we adopt, but the truth is that we are never given the choice of whether we want the technology or not, nor are we told what the price will be. No one ever gives us an option when new tools or new tech come along. No one ever says, "You may have this, but this is what it is going to cost you." Technology just happens, and we are left to deal with the consequences, both good and bad.

When social media first arose, no one explained to us just what the cost would be in terms of privacy. No one said, "Knowing this, do you want Google and Facebook and Amazon in your life or not?" Google and Amazon and Facebook did not come with an instruction manual. No one explained how they would work and what they would do to us. Google and Facebook and their ilk just appeared one day, and in short order, became the dominant forces in society. We never got to vote on whether we wanted Google in our lives or not.

But you do have a choice. You do have options.

There are alternatives to complete submission to the technology and its ramifications. If you understand what each new piece of technology is doing to you, you can decide whether you want it in your life or not. And even if you do want it in your life, once you understand what that technology is going to do to you, you can learn to control it, as opposed to it controlling you.

A TALE OF TWO TECHNOLOGIES

Long before Iran or Iraq or ISIS or Al Qaeda, there was an Islamic Revolution that shook the world. This one started in Sudan in 1882. Sudan had just become a British protectorate, a part of Egypt, which was also a British protectorate, and a part of the British Empire. In those days, Britain ruled about one quarter of the planet. The Mahdi, the title of a religious zealot along the lines of Bin Laden, had taken control of Sudan. He had also threatened the small British population that was then living in Khartoum, the capital. General Gordon, a British commander, led a force attempting to evacuate a British-friendly Egyptian forces from the city.

The British sent General Sir Herbert Kitchener and a force of eight thousand British soldiers to Sudan to help. Kitchener and his troops arrived two days later, too late to save Gordon, who was killed in Jannuary 1885 by rebels breaking into the city, apparently against the Mahdi's wishes. The Mahdi died later that year, and infighting ensued among possible new leaders. Eventually, sensing an opportunity, the British returned to the Sudan in 1896. On September 2, 1898, Kitchener and his new forces encountered the new Mahdi and his army of fifty thousand fanatical supporters in what is called the Battle of Omdurman.

The battle did not take long. It began at 6:30 a.m. and ended by 11:00 a.m. When it was over, there were twenty-five thousand Mahdist casualties, with five thousand captured. The British lost forty-seven. It was a smashing victory for the British Empire because the British had something that the Mahdists didn't have. The Mahdi's forces may have had religious fervor and nationalism on their side. The British had a new piece of technology—the Maxim gun, the world's first machine gun. As British historian Hillaire Belloc wrote:

> Whatever happens, we have got
> The Maxim gun, and they have not.

Technology won the day.

There were two very interested observers who had joined up with Kitchener on his expedition to Sudan. One was the twenty-five-year-old Lt. Winston Churchill, working as a journalist for the *Morning Post* and covering the war for the newspaper readers back home. The other was Captain Adolf von Tiedemann, a German observer invited by Kitchener to watch the Battle of Omdurman. Both he and Churchill took extensive notes.

On July 1, 1916, some eighteen years after the Battle of Omdurman, 1,530,000 British army soldiers accompanied by 1,440,000 French soldiers faced 1,500,000 German soldiers at the Battle of the Somme, in France. By November 18, when the battle was over, there would be

456,000 British casualties, 200,000 French casualties, and 500,000 German casualties—a total of 1,156,000 casualties from one battle.

What the British had done to the Mahdi's forces in Sudan the Germans were doing to the British and the French; by the same token, the French and the British were doing it to the Germans—mechanized slaughter.

Was anyone surprised? They should not have been. They had all seen what the new technology could do. And yet, the mechanical slaughter went on day after day, year after year, until an entire generation of Europeans were more or less exterminated by a machine.

At the Somme, and prior to it, British and German and French military doctrine had been to march toward the opposing army. The idea was to gain as much ground as they could and to push the opposing army into retreat. This technique had worked well since the time of Alexander the Great and up to and through the Napoleonic Wars.

The problem was the arrival of the new technology. Every time the British or the French or the Germans came out of their trenches and started heading for the enemy lines, the enemy simply began mowing the advancing soldiers down like so many blades of grass on a summer day. Unfortunately, the people running the British army and the French army and the Germany army simply could not think of any other way to fight a war. So they kept throwing young men up and over the parapet to run directly into the never-ending machine-gun fire. As it turned out, the British secretary of state for war was none other than General Sir Herbert Kitchener, the hero of Omdurman.

When the Battle of the Somme was finally over, the British forces had advanced some six miles, at a cost of 656,000 British and French casualties. Clearly, something had gone terribly wrong. A new technology had arrived that no one really fully understood or, rather, wanted to understand. No one had told the people of France and England and Germany (for it was they who would do the fighting and the dying), "OK, you may go to war, but this is the technology and this is what is going to happen to you and your children. It's your choice."

Perhaps had the public been educated about what a machine gun could do, the British and the French and the German public could have said, "No, we are not doing this." An educated public does not have to be led like sheep to the slaughter by the machines, nor do they have to be. But the machine gun and warfare were not considered the business of the public. It was the purview of the military. The public had no business getting involved in, commenting on, or, worst of all, having a role in how this new technology might be used.

Perhaps before Mark Zuckerberg unleashed Facebook on the world, he might have been made to sit down and explain to all of us exactly how Facebook was going to work and what it was going to do to us. Then, in full knowledge, we might all have been able to say, "No thank you," or perhaps, "OK, we'll take it, but you have to build in these guarantees." You can't open a restaurant and start serving food to people without an inspection that guarantees that the food is not going to kill you. Perhaps we need the same kind of guarantees before someone is allowed to unleash a new technology upon us.

This is of particular import as we begin to play with things like artificial intelligence and gene manipulation. These are the Maxim guns of the future, and we should take command and control of them before it is too late.

A DIFFERENT APPROACH

Some five hundred years ago, what was perhaps the most disruptive piece of new technology was unleashed. Yet its consequences proved far more beneficial than destructive, and that was because, unlike the Maxim gun, average people took control of that tech, almost from its inception.

In 1439, Johannes Gutenberg, mediocre goldsmith, blacksmith, and erstwhile entrepreneur at the very start of capitalism, was on the verge of personal bankruptcy.

Trying to get into business for himself, he had an idea to make a fast buck by manufacturing and selling "holy mirrors" that were supposed to capture the reflected light from religious relics and hold them for the owner. Think of it as a kind of salvation battery—it would store salvation until people needed it. It had been Gutenberg's plan to sell these holy mirrors to pilgrims coming to Aachen, in Germany, for a religious festival. The sale of religious trinkets was a big business in the fifteenth century, and Gutenberg, sure he was on to a good thing, had borrowed money to pay for the manufacture of his holy mirrors. However, unfortunately for Gutenberg, a flood in Aachen caused the festival to be cancelled, leaving Gutenberg with a big pile of holy mirrors and an even bigger pile of personal debt.

As there was no Chapter 11 in the fifteenth century, and needing to repay the loans that had financed his holy mirror business, Gutenberg, ever the entrepreneur, had one more idea up his sleeve. It would turn out to be an idea that would turn the entire world upside down.

Gutenberg hit on the idea of using his goldsmith skills to craft individual letters that could then be assembled, placed in a wine press, coated with ink, and used to print the pages of the Bible, the most popular book of its time. Up until then, all Bibles, and in fact the few books that existed, had been laboriously written by hand, one copy at a time. This was tedious and time-consuming, making each book incredibly expensive and rare. In 2007, Philip Patterson, who lives in Massachusetts, set out to copy the entire Bible by hand, all 788,000 words of it. Patterson worked on it for fourteen hours a day. It took him four years to complete the project, and that was without drawing the illuminated illustrations that generally filled medieval Bibles.

Because books were so expensive and difficult to produce, most people never saw a Bible or any book at all in their entire lives. And of course, with almost no books in circulation, the ability to read or to write was about as rare and as useful as the ability to read and write Sanskrit.

Gutenberg saw a clear opening. If he could print and publish Bibles quickly and all by himself, he could undercut the monks and own the Bible business, such as it was. The printed Bible business proved to be a smashing success for Gutenberg. But in inventing the printing press, Gutenberg had unleashed a new technology that was going to go far beyond cheap Bibles.

The printing press made it possible for anyone with an idea to distribute that idea to the world at next to no cost. Prior to Gutenberg, the dissemination of ideas was limited to the number of those who people could personally talk to in their town or village, which, more often than not, could be counted on the fingers of one hand. Thus, the world of ideas was stillborn. No matter what idea someone might have had, no matter what scientific discovery someone might have made, no matter what advance in planting seeds or plowing the fields or new ways to cook or hunt or smelt metal someone might have come up with, that knowledge was, more or less, destined to be either completely lost or quite limited in its distribution.

YOUNG MAN LUTHER

One person who saw the potential of Gutenberg's new technology was Martin Luther. Luther was an insignificant monk in an insignificant town in Germany. But he was annoyed with the pope. You are familiar with the story of Luther nailing his ninety-five theses to the door of the cathedral at Wittenberg, which gave birth to the Protestant Reformation. Had Luther only nailed the theses to the door of the cathedral, very few people would ever have read them. But Luther took control of the new technology himself. He started printing and distributing his ideas, using the press to effectively say no to the Church and the monarchy. By publishing, Luther was able to distribute his ideas to the world. As it turned out, Luther's ideas went viral, or at least as viral as the technology of 1517 allowed.

Nothing quite like that had ever happened before. Luther's ideas, of course, resonated with many people, but prior to Luther and his printing press, none of them would have had the reinforcement that came from seeing ideas in print with which they agreed. The printed word, something that had existed only in the Bible prior to Gutenberg, carried enormous resonance. Luther's ideas resonated in Europe, and soon people from England to Italy read what Luther had written and agreed with it. Luther continued to publish, and within a very few years, due to the popularity of his books—which he sold—Wittenberg went from being a small town to the publishing capital of the world.

In Rome, Pope Leo X was angered. Who was this insolent monk who was proving so very threatening to the established order of the world? Leo called upon the Emperor of the Holy Roman Empire, Charles V, to shut down the whole Luther operation. And so, on

April 18, 1521, Luther, as ordered, appeared before Charles V in the city of Worms on the Rhine.

The Diet of Worms was a formal accusation of Luther for acts of heresy. Luther had already been excommunicated from the Church on January 3, 1521. But worse was in the offing. The Czech dissident Jan Hus had been burned alive at the stake in 1415 for similar transgressions. But Hus was not a published author with a substantial following. This was something totally new. Thanks to the new technology, Luther already had many followers.

The Church could have executed Luther, but it did not. The hierarchy knew already that to do that would only elevate his movement and turn him into a martyr. It was, in many ways, the first incidence of mass media and all of its ramifications, a phenomenon with which we are all quite familiar today, but an event that was completely alien to the inhabitants of the fifteenth century. As the Maxim gun had been the exclusive purview of the military, the printing press and its power to mass-distribute the Word might have also remained in the hands of the Church and the monarchy. But Luther took control of the technology and used it for his own ends as opposed to those of the Vatican and the state, and that made all the difference. He had democratized the enormous power of that new tech and, in doing so, opened the door to the ultimate ramifications of that democratization, which would ultimately include, but not be limited to, the Enlightenment, the Scientific Revolution, the First Amendment, books, literature, magazines, newspapers, free thought, free speech, and the rise of western liberal democracies.

LUTHER WAS NOT ALONE

In 1010 AD, Thorfinn Karlsefni discovered America. He was a Norse Viking, and he sailed from Greenland to what he called Vinland, and what we today call Newfoundland, and founded the first European settlements there. You would think that was a pretty big achievement, that the United States of America should more properly be called the United States of Karlsefnia. You would think that there should be a Karlsefni University in New York and a Karlsefni, Ohio. But there isn't. There isn't even a Karlsefni Day Parade. Even though Thorfinn

Karlsefni preceded Christopher Columbus by some five hundred years, he gets no recognition. How come? Despite the great achievement of Thorfinn Karlsefni, he left out the most important part of discovery and exploration. He forgot to tell anyone.

Karlsefini's achievement occurred five hundred years not just before Columbus but also before the invention of the printing press. Columbus's discovery, in 1492, was about forty years after the printing press, and was one of the first great topics publicized via this new technology. When Columbus returned to Spain, he used the printing press to publish a letter he had written explaining the voyage and what he had discovered. Columbus may not have been the first European to discover the New World, but he was certainly the world's first publicist. Nearly three thousand copies of his first letter were printed and distributed across Europe. Had Ferdinand and Isabella, who had underwritten the cost of the voyage, understood the power that this new technology would bestow, they doubtless would have secured it solely for themselves. The New World had untold riches, and Isabella and Ferdinand, having paid for the trip, should have wanted to hold on to a monopoly on the IP of the discovery. They didn't.

Because Columbus published, everyone was soon busy reading about what Columbus had done and had found. It was not the discovery nor the voyage per se that had such a massive impact on what was to come, but, rather, it was the marriage of that achievement with the new technology of the printing press in the hands of Columbus.

Prior to the printing press, the only way someone could have learned what Columbus had done would have been to come to Seville and meet Columbus in person. But with the printed word now in circulation, all they had to do was to read what Columbus had done, and say, "If he can do that, so can I." And so the whole great enterprise of European exploration and conquest of the New World began—not by Columbus, per se, but driven by the unforeseen impact of the printing press.

You might ask why we are not called the United States of Columbia. This too was a product of the printing press and its unintended consequences.

In 1507, Martin Waldseemüller, a German mapmaker, published what was the first world atlas (the nature of the world having been enormously changed by Columbus' own discovery and letter). As with Columbus's letter a book of maps was an entirely new concept. Waldseemüller's book, *Universalis Cosmographia,* contained the first maps of the New World.

Amerigo Vespucci, inspired by what Columbus has done (or, as some say, inspired to break the Columbus monopoly on the New World), in April 1495, was licensed by the Crown of Castille to also explore the New World, and he made it to the West Indies. As Columbus had done, Vespucci published upon his return his own accounts of his voyages. It was this book that Waldseemüller was believed to have read, and that which motivated him to name the New World "America" for Vespucci, as opposed to Columbia for Columbus. That's the power of the press.

Vespucci, as it turned out, was something less than a savory individual. He had been a pimp in Florence, later a ship's chandler, and as such, he supplied people such as Columbus with pickled fish and meat for the long voyages. In the nineteenth century, American author Ralph Waldo Emerson wrote the following:

> Strange, that the New World should have no better luck,—that broad America must wear the name of a thief. Amerigo Vespucci, the pickle-dealer at Seville, who went out, in 1499, a subaltern with Hojeda, and whose highest naval rank was boatswain's mate in an expedition that never sailed, managed in this lying world to supplant Columbus, and baptize half the earth with his own dishonest name.

The arrival of the printing press and, more importantly, its democratization, essentially turned the entire world upside down. It marked the end of the Middle Ages and the beginning of the Enlightenment, the Reformation, the intellectual and scientific revolution, and so much more. The printing press democratized the medium of print, placing the power and the ability to write, to publish, to share ideas, to participate in a national and later global dialogue, to open the hitherto closed world of politics and public discourse to everyone. It was an earth-shattering moment in human history. But it was only so because the public effectively seized control of the technology from almost its inception. It would have been a far different and far less interesting story had the tech remained in the hands of the powerful few.

The Constitution of the United States, as it turns out, is a direct product of the Gutenberg revolution. A printed document that clearly outlined the delineated the power and, more important, the limitations of power of the new government, printed and published so that everyone could see it and read it. The First Amendment of that document does not say, "You have the right to vote"; it does not say, "You have the right to representation." It says, "Congress shall make no law abridging a free press." The founders who wrote and published that document understood that the very foundation of a free society was based on the concept of a free press.

We no longer live in a world of print. For better or worse, we now live in a world of media and video. Up until now, that media has been owned and controlled by a small handful of very powerful corporations. Until now, there was no "free press" in the realm of television or video. The cost and complications and difficulty of creating and producing content in that medium, not to mention distributing it, made it all but impossible for anyone except the extremely rich and the extremely powerful to get their hands on either the means of production or the means of distribution. Today, in America, fully 90 percent of what you see or hear is controlled by six major corporations—they are the Catholic Church and the monarchy of the twenty-first century.

There is no rule that says that new technology must reside in the control of the existing power structure or the government. This applies not just to modern media but also to the rise of artificial intelligence, gene manipulation, and hosts of other new and powerful technologies on the horizon. With the arrival of each new technology, we are all faced with the choice of acceding to the technology and all its implications or taking control of the technology and saying "no, not for me." That choice is always in your hands, but the powerful do not surrender power easily.

6

The Machine That Ate the World

On the evening of December 28, 1895, a small and select group of people were invited to an unusual event in Paris. It was to be held in the basement of the popular Grand Café on Boulevard des Capucines in a room called Salon Indien. The invited didn't know each other, nor did they know exactly what they had been invited to attend. They had only been told that it would be "interesting."

There was a lot that was interesting going on in Paris at the end of the nineteenth century. Paris was the arts and culture capital of the world. The city had been the witness to radical revolutions in music, in dance, in literature, in theater, and in painting and sculpture. Now it would be a witness to one of the most radical revolutions of the past five hundred years, but this time in media, a format that, until then, did not even exist.

As the invited filed into the windowless salon, they must have wondered just what kind of a show they had been invited to see. A ballet? A new play? A Debussy concert? There was no stage. There were no musicians. Even a display of new and cutting-edge paintings would have been at a gallery. The walls were bare. It was all unknown. The only thing they had been told was that the whole thing would only last about ten minutes, and then they could be on their way.

Many thought it might have all been a mighty waste of time and were anxious to leave before the performance even began.

Slowly, the lights went down and the group found themselves sitting in the dark.

Then, a flickering image appeared on the screen. It was a picture. They had seen this before, a magic-lantern show—photographs projected on a screen. If this was the "big amazing new thing" they had been invited to witness, they would be sorely disappointed. But wait! The pictures were moving! Was this even possible? What they saw on the screen was moving images of workers leaving the Lumiere Factory in Paris, portrayed in two dimensions, but many meters high. This was truly amazing. This really was magic. This was like a doorway to another world. This was the world's first motion picture ever, courtesy of the Lumiere brothers.

The Lumiere brothers, Auguste Marie and Louis Jean, were born in the small French village of Besançon to Charles-Antoine and Jeanne Josephine Lumiere in 1862 and 1864, respectively. Their father ran a small photographic studio, putting him on the cutting edge of visual technology, at least for the mid-nineteenth century. They moved to Lyon, where Charles-Antoine, whose photographic work was dependent upon chemically treated metal plates, came quite close to bankruptcy, until the brothers devised a mechanical system for processing the plates called *etiquettes bleue*, which not only saved the business, but expanded it sufficiently to allow him to ultimately employ a dozen workers. When their father retired in 1892, the brothers took over the factory and began to experiment with moving pictures.

The idea that it was possible to create the illusion of moving pictures is actually quite old in human history. It is all based on a phenomenon called "persistence of memory." When your eye sees an image, your brain records it just a bit longer than it is actually there. It lingers in your memory for a second or two. If, in that time, you can join it to another similar image, the two seem to flow together. This is how we see movement.

An earthenware bowl, excavated by archaeologists in present-day Iran, and dating from some three thousand years ago, would seem to be the first example of the use of this persistence of memory to create the illusion of motion. The bowl is decorated with a series of painted images of a goat jumping and eating leaves from a tree. If you spin the bowl fast enough, apparently, you get what was most likely the world's very first "motion picture."

The first modern manifestation of this phenomenon was carried out by an eighteen-year-old sophomore at Brown University in 1865. William Ensign Lincoln built the first working zoetrope. It was essentially a spinning wheel inside a casement. When someone spun the wheel and looked through a small opening, the person saw the illusion of motion. Lincoln was smart enough to patent the idea (Brown is not Ivy League for nothing) and even smarter to sell the whole thing to the Milton Bradley Company, the same people who would one day go on to produce such well-known diversions as Chutes and Ladders and Candy Land. The company also produced the lesser well-known Happy Days in Old New England and Word Gardening. You can't win them all.

But with the zoetrope, it had a winner. The device proved enormously popular, and Milton Bradley patented and released it in Britain just a few years later. But the zoetrope had its limitations. Only one person could see the moving image at a time, and, as it was just a continuing rotation of the same paper strips, that story content was rather limited.

What the Lumiere brothers then did was to take the concept of the zoetrope and marry it to projection. The secret to their success was in their invention of perforated holes on a film strip, married to a gate and fork movement action that effectively stopped the image for a moment, then grabbed the next one and placed it to be projected. The brothers used photographs instead of paper images.

The Lumieres shot on film that was seventeen meters long, which, when hand-cranked through the camera and later the projector, meant each movie lasted for about fifty seconds.

They called it "moving postcards," and it was an immediate smash hit. The brothers took their invention and their show on the road and opened in Brussels on March 1, 1896. From there, they moved on to Bombay, Montreal, New York, London, and Buenos Aires, electrifying audiences wherever they went. The whole world pretty much got it immediately. Alas, the brothers did not. They saw it as a kind of toy. They said, "Cinema is a medium without any future." Instead, they went back to the lab and invented color photography, the *autochrome lumiere* process, and became major producers of photographic products. The Lumiere company was ultimately sold to Ilford, a major European photo business, and the brothers pretty much disappeared from history, but not before they had unleashed the flashing-lights virus on the world.

When the Lumiere brothers opened at the Café des Capucines, it was the first time anyone had ever gone to the movies, the first time any human being had ever seen a movie. The concept of a projected image simply had never existed before. Nine more short clips followed, each one less than a minute long. The brothers decided to keep showing their moving pictures in the Café basement, night after night, and charging one franc each. or about $12.50 in today's money. Word of mouth spread, and within only a few weeks, the queue outside the basement door stretched to a half-mile or more.

The attraction to seeing the movies was both immediate and incredibly powerful. The brothers had tapped into something very fundamental, something that seemed to have an overpowering attraction. They had discovered an addictive drug. The "movies" that first French audience saw on a cold December night in Paris would be considered unwatchable by us today. The films were about a minute long each. They were, of course, in black and white. There was no sound. There also was no editing. There were no scripts, no plots, no story lines. The films were made by just a fixed camera, capturing a single scene: a gardener using a sprinkler, Auguste Lumiere and his family having breakfast, his daughter trying to catch a fish.

Yet within weeks, tens of thousands of people were lining up, waiting hours on end, all for the same ten-minute experience. The brothers had reawakened a long dormant strand of DNA that we all carried with us, without knowing it. They had also, again inadvertently, created a mechanism to marry the very deep and ancient tradition of storytelling to our addiction to flashing lights. It would prove to be a killer combination.

The brothers' discovery of this hitherto unknown aspect of our nature did not remain a secret for very long. It soon became obvious that moving pictures were destined to be a big business, despite what the brothers thought. Edison's newly invented movie camera and movie projector created the technological foundation for the birth of an entirely new industry—movies. And it was not long before Britain, France, Germany, Russia, and the United States began cranking out moving pictures by the thousands. There was a ready, built-in audience that would pay, in advance no less, to be bathed in flashing lights and stories.

The Lumiere brothers' discovery would, within only a few short years, give rise to Hollywood and ultimately the entire global motion-picture industry. Soon, it would be completely normal to spend your Sunday afternoons sitting in a darkened room, staring at a screen. But this would only mark the very beginnings of what would ultimately become our number one obsession, eclipsing everything else. The seed had been planted; the virus had been unleashed.

THE BIRTH OF TELEVISION

On September 8, 1913, a tiny article appeared in the *New York Times*, buried way in the back of the paper. It stated that an obscure discussion had gone on in an esoteric physics lecture in Birmingham, England, the previous week. Sir Oliver Lodge, a British physicist, had taken issue with a rather radical theory put forth by a hitherto unknown German scientist named Albert Einstein. The radical theory was something called "relativity." Lodge thought it was nonsense. That the *Times* even covered such a story bears witness to just what "the newspaper of record" meant in those days.

Buried on the back pages of the paper, covering an obscure subject, the tiny article no doubt was little noticed by most people. Yet, even if they had read it, hardly anyone would have understood the earth-shattering import of that bit of print. Yet there it was—a moment, frozen in time, that would change all of human history. Embedded in that article was news, in a sense, of the coming of the atom bomb and all the weight it carried for changing the world and the possible destruction of humanity. One would have thought that the newspaper might have given that one the front page. But it didn't. The front page, on that day, actually carried the seemingly far more newsworthy story that Champ Clark had been reelected Speaker of the House. Today, no one knows who Champ Clark was, but there is hardly anyone who has not at least heard of Einstein or his theory of relativity, even if they can't explain it.

In 1927, one of the greatest changes ever to confront humanity was also reported in the *New York Times*. But at the time, like the birth of relativity, hardly anyone took any notice of it. On April 8, 1927, the *Times* published a small story headlined, "Far Off Speakers Seen as Well as Heard Here in a Test of Television." Herbert Hoover, later to become president of the United States, but at that time Secretary of Commerce, had given a speech in Washington, DC, that had remarkably been transmitted more than two hundred miles away to a waiting receiver in New York. What had been transmitted was both sound *and* pictures. Remember that radio had barely come into its own as the first electronic medium in the 1920s. Yet at that moment, television had officially been born. Ironically, barely anyone paid any attention.

Of course, the television that Hoover appeared on was a good deal different from the seventy-two-inch model that you probably have hanging on your living room wall. First, the image was black and white. Second, the screen measured only a few inches square, about half the size of a postcard. Third, and ultimately the most important, the image was, frankly, hard to see. It was extremely grainy and the resolution was terrible.

Despite its small size, however, the experiment, which was seen more as a kind of scientific curiosity, was deemed a great success. "Like a Photograph Come to Life," read the *Times'* headline, almost a memory of the Lumiere Brothers' moving postcards. The story continued on the jump to a full page, which was quite a commitment from the *Times* for what was essentially a story about a massive engineering experiment. This experiment had been carried out by Bell Labs, the Whippany, New Jersey–based research arm of AT&T, a part of the Bell Telephone Company. Bell Labs would ultimately go on to confirm the big bang theory and invent the laser, the transistor, the CCD (charged coupling device), computer operating systems, Unix, the basics of computer programming, and much more.

The concept of television had been, until Bell Labs put the thing together, almost entirely theoretical, at least in the United States. Secretary Hoover was a good sport to allow the scientist and technicians from Bell Labs to install and run all the expensive, complex, and probably dangerous equipment necessary to capture his image and send it up to New York City. Even though the *Times* extolled the technology and the experiment was deemed a smashing success, a subheading on the page noted, "Commercial Use in Doubt." Wrote the *Times*, "The commercial future of television, if it has one, is thought to be largely in public entertainment—super-news reels flashed before audiences at the moment of occurrence, together with dramatic and musical acts shot on the ether waves in sound and picture at the instant they are taking place in the studio."

The successful experiment, however, went nowhere. Having proven that television was at least theoretically possible, AT&T soon abandoned the entire project. The problem beyond the monetization issue was that the image was just terrible, and AT&T, for all its technical wherewithal, could see no way to improve it.

GERMANY GOES FIRST

Despite the *Times* article, AT&T was not the first to attempt to build a television machine. The credit for that must go to a Prussian/ German named Paul Nipkow, who in 1885 received a patent from the German patent office for an invention he called the Nipkow disk.

In its simplest explanation, the Nipkow disk was a series of spinning disks with holes in them that allowed light to pass through. By breaking an image into bits of light, via the spinning holes, Nipkow thought it would be theoretically possible to transmit images through a wire or over the air. The images would then be picked up by a receiver, which was also a series of spinning discs. On the receiving end, the process would effectively be reversed and the fragmented light images would be reassembled, and thus the image and its movement would be replicated. It was, in a sense, a very sophisticated version of flipping though a stack of photos of a golfer making his swing. What the relatively new telephone had done for sound, the Nipkow disk would, in theory, be able to do for pictures.

Even though Nipkow conceived of the concept of the spinning disks and received a patent for it, he never built a working model. The credit for that would have to go a Scotsman named John Logie Baird, who on January 26, 1926, working from Nipkow's plans, built the world's first working model and carried out a demonstration of the world's first "television" for members of the Royal Institution and a reporter from the *Times* of London. The Nipkow disk was also the basis of AT&T's working demonstration a year later. Using Nipkow's spinning disk concept, Baird was able to build a television system that could actually transmit images made of thirty lines of resolution.

However, even the ability to transmit and then decode thirty lines of resolution was quite an achievement. A few months earlier, working out of his Soho lab in London, Baird had propped up a dummy named Stooky Bill in front of his camera and switched on his apparatus. As the disc began to spin, Baird was able to transmit the images of Stooky Bill at five frames per second and thirty lines of resolution. Contemporary television would ultimately transmit at thirty frames per second and a resolution of 1,088 lines, for purposes of comparison.

Although Baird's images may have been rough, they were real. The machine actually worked. Anxious to see what a human would look like on TV, Baird ran downstairs, and grabbed an office worker, twenty-year-old William Taynton, who was propped up in the studio in front of the camera, replacing Stooky Bill, and thus became the first person ever to appear on TV.

Baird had achieved what had been a human dream for years—the ability to transmit and to see moving pictures from somewhere else. He had made the world's first working television set. Anxious to publicize his new invention, Baird rushed over to the offices of the *Daily Express* newspaper. In one of the great moments of journalism, the editor of the paper is reported to have said, "For God's sake, go down to reception and get rid of a lunatic who's down there. He says he's got a machine for seeing by wireless! Watch him—he may have a razor on him."

But Baird was not done. On July 3, 1928, using the spinning disks, Baird transmitted the world's first color television (or *colour*, as it is spelled in the United Kingdom).

Baird's achievements were so impressive that the BBC, which had until then been a strictly radio-based operation, adopted the Baird system and began making experimental television broadcasts.

But the Nipkow spinning disks, upon which both Baird and AT&T's work were based, had an inherent flaw. The picture quality was terrible, and it wasn't going to get much better. To make it work at all, the system required massive spinning disks, both at the point of transmission and the point of reception, to reconstruct the fractionalized images and reassemble the pictures. The spinning disks not only required tanks and hoses but were also incredibly noisy. As well, they could only produce small images. Even to make a four-inch square picture would require spinning disks six feet in diameter, something you probably would not want in your bedroom.

The failure of the Nipkow disks, after so many years of trying, seemed to indicate that television might have reached a kind of technical cul-de-sac. Maybe it wasn't possible. Or, maybe an entirely new approach to transmitting pictures and sound was needed. That turned out to be the case, but it ended up coming from the least expected of places.

A FARM BOY CHANGES THE WORLD

In 1843, Charles Babbage and Ada Lovelace designed and attempted to build what would, in retrospect, be the world's first computer. It was called an analytical engine, and while the concept was sound, the technology of the nineteenth century simply was not adequate to the task. It would not be until the Second World War, a full century in the future, that the first electronic computer, the ENIAC, could be built. What was required was a fundamental leap from mechanical to electronic design. The same transition was necessary to make television workable, and in doing so, to make it a viable technology to the world.

Philo T. Farnsworth was born on August 19, 1906, in a log cabin in Beaver, Utah, to Lewis Edwin Farnsworth and Serena Amanda Bastian. Lewis's father had built the log cabin that they lived in with his own hands. These people were poor—dirt poor. In 1918, the family moved to Rigby, Idaho, where Lewis supplemented his meager income as a farmer by hauling freight with his horse-drawn wagon. The Farnsworths were members of the Mormon Church.

The move to Idaho was momentous, but not in a way you might expect. The prior occupant of what was now the new Farnsworth homestead had left a pile of old electronics magazines in the attic, a collection that twelve-year-old Philo soon discovered and devoured.

Electronics was not terribly advanced in 1918. Unlike many American homes in Idaho at that time, the house had just been wired for electricity powered by a Delco generator that broke down often. Philo figured out how it worked, repaired it, and rewired and rebuilt the generator. He seemed to have a natural gift both for technology and for tinkering. He won a $25 prize offered in a magazine for his design for a magnetized car lock.

Magnetized car locks and repairing a generator are one thing; solving the riddle of how to create a workable electronic television system is quite another. According to legend, Philo was plowing his father's field one day with a horse-drawn plow when he noted the images that the endless parallel lines of wheat seemed to create on the rolling fields of the family farm. Could one not create, he wondered, images using lines like these? This insight, apparently, became the foundation of the scanning lines that made electronic television work.

Apocryphal or not, young Farnsworth approached his high school science teacher, a Mr. Justin Tolman, for advice on building an electronic television system. Tolman taught both chemistry and physics at Rigby High School, and Farnsworth, anxious to work out his concept, provided Tolman with drawings of his idea for a line-driven television system.

In 1922, Farnsworth matriculated at Brigham Young University in Utah, with the family moving to Provo, Utah, a year later. While at the university, he studied electronics and received a certificate from the National Radio Institute. It was at BYU that he met Elma Gardner, known as Pem, whom he would later marry. Farnsworth was forced to leave the university after the death of his father in 1924.

Farnsworth must have already been thinking ahead. He was accepted into the United States Naval Academy at Annapolis, receiving the second highest application scores in the academy's history.

However, after a short stint there he invoked a clause in the rules that allowed the eldest sons of recently deceased fathers to leave. The real reason was that he found out that any inventions or patents he created while at the academy would be the property of the US Navy.

Farnsworth had other ideas. He set up a radio-repair business with Pem's brother, Cliff, but it soon failed. It was here that fate intervened. Two investors, Leslie Gorell and George Everson, who found Farnsworth through a University of Utah jobs program, agreed to invest $6,000 in Farnsworth's fledgling idea for television, the equivalent of $90,000 today. With their money, Farnsworth married Pem and moved to Los Angeles to open shop and get to work on inventing television.

Philo Farnsworth was not alone in his fixation on the idea of creating a truly workable television system. Enter David Sarnoff.

AT&T FIGHTS BACK

David Sarnoff was born in 1891 in Uzlyany, Russia. For a sense of perspective, think of it as being born on the moon by today's standards—the far side of the moon. Sarnoff spent the first nine years of his life studying the Talmud and Torah at the local heder, or Hebrew school. Electronics were not high on the list of subjects at the heder. In fact, they were not on the list at all. In 1900, along with his family, David emigrated to the United States, landing in New York. He immediately went to work helping to support his family by selling newspapers and studying at night at the Educational Alliance. In 1906, David's father Abraham became incapacitated by tuberculosis, and David had to go to work full time to support the family.

Sarnoff wanted to be a journalist. On September 30, 1906, looking for a job, he walked into the lobby of the *New York Herald* newspaper and announced to the first person that he ran into that he wanted a job. Unbeknownst to Sarnoff, the man he had approached worked for the Commercial Cable Company, which leased office space from the *Herald*. The cable company was looking for a messenger boy. Thus does history turn on such small accidents. Sarnoff was fifteen years old.

Sarnoff proved himself to be a model employee, studying to learn

Morse code at night. However, he was soon fired for taking off Jewish holidays but was able to parlay his experience and knowledge of Morse code into getting a job with the Marconi Company.

In 1898, eight years earlier, Guglielmo Marconi had shattered the world by inventing radio. Before radio came along, the world's dominant means of fast communication was the telephone. Invented by Alexander Graham Bell in 1873, the telephone had turned the world on its head. Prior to the invention of the phone, the fastest anyone could get a message to anyone else was by the telegraph, whose syntax was limited to dots and dashes, which took both time and skill. Prior to that, jumping on a fast horse and galloping was the fastest was to deliver a message. Thus it had been since the days of the Roman Empire, and before that. The telephone changed everything, and the notion of the telephone captured the imagination of the world.

The invention of the telephone did more than allow people to talk to one another over great distances. It also created a fundamental change in the whole notion of value, and in some ways, that proved more significant than the instrument itself. It would be a transformation in some very basic assumptions of how things were supposed to work that would one day culminate in the Internet. Prior to the invention of the telephone, value was equated with rarity. If you had the only diamond in the world, it was worth a great deal. If, however, suddenly everyone was to have a diamond, well, then your diamond was worth a lot less. People prized rarity; they horded it; they protected it at all costs. People locked their treasures up in a safe or a treasure chest. They tried to corner the market on commodities like copper or silver or, later, oil.

The telephone changed everything. If someone had the only telephone in the world, you might think it would be worth a lot, but you would be wrong. Having the only telephone in the world is, in fact, worthless, as there would be no one to call. The value of a phone is directly related to the number of other phones in the world. If there are only two phones, then you can only call one other person. If there are a hundred phones, then suddenly your phone is capable of calling one hundred people. A thousand is better, a million better still.

It did not take long for the financial world of the mid-nineteenth century to understand that the old rules had been turned upside down. For the first time in human history, value was dependent not upon rarity, but rather upon commonality. The more phones there were, the more common they were, the cheaper they were, the easier they were to get and install, the more valuable the phone became. It was the complete opposite of thousands of years of human experience.

As a result, the initial investors in Bell's new company, Bell Telephone, could quickly see that their investment was going nowhere unless everyone had a phone. If everyone had a phone, then, in fact, no one could afford not to have one. Value was suddenly tied to making phones available to everyone. It was the world's first network.

The invention of the phone was a great technological achievement, but what was really needed now was engineering—perhaps the biggest engineering project that the world had ever attempted. The building of the Brooklyn Bridge was nothing compared to wiring an entire nation. In fact, if this phone thing was to work out, it would be necessary to wire up every store, every business, every home in America, if not the world.

Today, we live in a wired world. We take it for granted that every building we walk into is wired for electricity, for phones, for cable TV, and, increasingly, for broadband. But in 1873, this was not the case. No one was wired. Nothing was wired.

Building a working telephone that could connect two offices was one thing, but wiring the world was another. Wiring the world was going to cost money, so Bell Telephone sold shares and bonds and loans were taken, but to everyone involved, wiring the world seemed like a no-brainer investment. It was sure to pay off. Who would not want to be wired in? Who would not want a phone? The phone and, more significant, the wires that supported the phone, for which people would have to pay, was going to be the Internet Revolution of the nineteenth century. So people invested in droves. And those investments brought results.

By the turn of the century, most of the country and a good deal of the world was well on its way to being wired. It was a stellar, albeit vastly expensive, achievement. But it would seem to have been money very well invested. It was only a matter of time before all those millions of miles of wire, not to mention the poles that they were strung from, the connections, the relays, and the electrification of it all would pay off.

Then, in 1898, Guglielmo Marconi, an Italian inventor and electrical engineer, seemingly out of nowhere, arrived with his new invention: a black box. "Watch this!" Marconi might have said. "I can talk to him on my new machine without wires! I call this . . . wireless!"

Wireless! What a great idea—unless you had just invested your life savings in wires. Then it was a terrible idea.

So AT&T, a company Bell had established in 1889, did what any American corporation would do. It bought the rights to the audion, the key piece of tech that made radio work. And it sought to suppress radio's development. Because the world of the 1890s was essentially telephone-centric, AT&T and Bell could only see radio as a kind of "wireless telephone." Ever hear the term "radio-telephone"? That's what people thought radio was for.

With the audion patent in the hands of AT&T, the only place that the Marconi Company could develop radio and not compete with the telephone and its wires was on ships. Ships were fine from AT&T's point of view, because it didn't really work to wire up ships. When Sarnoff walked into the offices of the Marconi Company, despite its having the rights to what you would think would have been a multi-billion-dollar piece of technology in radio, it was a relatively small company. Owning the technology, it turns out, is not everything. What a company does with the technology counts.

The Marconi Company's primary clients for its wireless gear were ships and shipping lines. Ships at that time were just beginning to experiment with wireless as a way to stay in touch with the home company, and soon many ships, both navy and commercial vessels, began to carry radio as a kind of experimental accessory.

THE *TITANIC* GOES DOWN WHILE RCA GOES UP

At 11:40 p.m. on the evening of April 14, 1912, the *Titanic*, a passenger liner operated by the British White Star Line, and at that time, the largest ship afloat, struck an iceberg in the North Atlantic. The collision opened five of the ship's sixteen waterproof compartments, just enough to cause her inevitably to sink. Carrying 2,240 passengers and crew, but only twenty lifeboats, with a total capacity for 1,178 people, it was a disaster waiting to happen. And on the evening of April 14, it did.

Along with its limited supply of lifeboats, the *Titanic* also carried radio, supplied and manned by the Marconi International Marine Communication Company. The Marconi Company had equipped the Titanic with what was then the most powerful radio transmitter in the world, capable of sending its radio signal nearly four hundred miles. The radiomen, Jack Phillips and Harold Bride, who were on the *Titanic* didn't work for White Star Lines; they were employees of the Marconi Company.

As the ship began to sink, both Phillips and Bride began to send out radio distress calls. Some four hundred miles away, the then twenty-one-year-old David Sarnoff was manning the Marconi radio station at the Wanamaker Department Store in Manhattan. Sarnoff was the first one to pick up the distress signal from the sinking *Titanic*.

Today, we are used to seeing events live, as they happen. When the airplanes struck the World Trade Center on 9/11, the whole world watched live. When the US struck Iraq with its "shock and awe," the whole thing was broadcast live for a prime-time audience. In fact, when any disaster strikes, the first thing you do is turn on the TV (or maybe now check out your Twitter feed), to see what is happening, as it happens, in real time. For us, this is a given.

But when the *Titanic* struck its iceberg in the mid-Atlantic, the world was vastly different. News events could take days to reach the public. That was the norm. In the case of the *Titanic*, however, because of radio and its instant transmission, the world was able to "watch" the tragedy unfold in real time, or at least follow it, as it happened. Nothing like that had ever happened before. It immediately captured the attention of the nation.

In the past, when there was a "breaking news event" (though the term didn't yet exist), crowds would gather in Times Square around the *New York Times* building, where copies of the latest edition of the paper would be "hot off the press." The famous news zipper, the electronic lights that circled the top of the building and electrified the headlines for all to read, would not be added until 1928.

But with the *Titanic*, newspapers as the primary point for news and information had suddenly been eclipsed by this new technology of radio. For seventy-two hours, Sarnoff stayed by his post, relaying updated bulletins of news from the sinking *Titanic*, and later, the names of survivors plucked from the freezing water by the *Carpathia*, which also had responded to the *Titanic*'s radio distress calls.

Crowds in New York did not gather in Times Square as they would have earlier, but now gathered outside Wanamaker's Department Store, awaiting each update from young Sarnoff. At that moment, David Sarnoff had an insight. Radio was not a competitor for telephones. In fact, radio had nothing to do with telephones or person-to-person communications at all, which was the basis of telephony. Telephones were one person talking to another. Radio was about broadcasting a single voice to thousands, perhaps one day even millions of people, all at the same time. At that moment, David Sarnoff invented broadcasting. It was an idea that would completely change the world. Up until then, owning a radio at home had also meant owning a transmission and reception station. It was remarkably complex and expensive. It was also, in its own way, interactive. At this very primitive stage of radio, it was looked at as a participatory medium; that is, anyone could create and transmit radio as well as receive it. Radio owners talked to each other and expected a response.

But Sarnoff envisioned something entirely different, a simple device he called a radio music box that would have only two controls—volume and frequency. The key factor to Sarnoff's invention was that it was passive. That is, that radio owners would only be able to receive what Sarnoff decided to transmit. And thus, radio and in a larger sense, broadcasting as we understand it today, were born.

Prior to radio, the best and in fact the only real source of mass media was newspapers, a child of the printing press. But newspapers had an inherent limit to their effectiveness and their reach. They had to be physically printed—one copy for each person who received the content. This was expensive. With radio, all you had to do was create one version and suddenly the content was infinitely repeatable. Each newspaper had not only to be physically printed but also physically delivered into the hands of each reader. Radio immediately eliminated this cost and barrier to distribution.

The issue, then, was how to make radio into a profitable business. The model for the newspaper business was old and established. You paid for each copy of the paper and you collected money from advertisers. Newspapers had two very healthy revenue streams. How could radio, which was free over the air, be monetized?

In 1919, seven years after the sinking of the *Titanic* and Sarnoff's initial moment of insight, radio was still far from a sure bet. During the First World War, as in all wars, technology was sped up. The War Department began to see the potential of radio. This would help grow the medium.

During the American Civil War, which had only been a bit more than one generation removed from the start of the First World War, electronic technology had come to the battlefield in the form of the newly invented telegraph. Lincoln, who had grown up in a log cabin with no electricity, became mesmerized by the telegraph and would spend hours in his telegraph room, receiving information from his generals and relaying instructions. The Union had a vastly more sophisticated telegraphy network than the Confederacy, even if the Confederacy had much better generals. Lincoln was able to coordinate a war that stretched from Vicksburg to Virginia, and all from Washington. Nothing like that had ever happened before.

By the time the First World War rolled around, the ability to communicate with forces in the field, to move them based on real-time events, and to receive up-to-the-minute information from the front lines was proving critical. Land-laid wires, used both for telegraphy and telephony, were too easily cut, either by the enemy or simply by the never-ending explosions and shrapnel. Radio bypassed all that, and in 1917, with the entry of the United States into the First World War, the government either shut down or took over all private radio stations or operations. Radio became a tool of the military and the government. In fact, it was considered an act of treason for any private citizen to operate or even possess a radio transmitter or receiver.

A great deal of this was driven by a fear of radio espionage, which was pretty much baseless, but new technologies are often misunderstood. The potential for radio in the field was not misunderstood, and as quickly as the US government shut down private radio, it also took control of all the patents related to radio manufacture in the United States. These patents and the manufacturing of any and all radio equipment was placed in the hands of the military, and the War and Navy Departments maintained a monopoly on all radio operations until the end of the war in 1918.

Having effectively brought together all the patents and all the radio operations in the country, the US government then turned to General Electric and asked it to create a kind of national radio organization through which the Army and Navy could maintain their monopoly on over-the-air transmission. Thus was RCA, the Radio Corporation of America, born. Marconi Wireless was brought into the RCA amalgam, and thus did David Sarnoff become its first General Manager.

In 1919 radio was still more of a curiosity than anything else, and as the core patents that RCA held were related to the manufacture of radios, it seemed only logical that the business that RCA should be in would be selling radios. However, if people were going to buy radios, then there had to be something for them to listen to. It became clear quite early on that the only revenue stream for radio, aside from selling units, would be advertising. Advertising was and remains intimately linked to being able to garner and hold an audience, and garnering and holding an audience is intimately linked to attracting and holding it with the content. Hence, the content of radio was looked upon merely as the reason that people would want to buy radios to begin with. It was a bit like the razor business: the money was in the blades. But what would the content be?

WHAT IS THE MEDIA FOR?

In 1920, no one really knew what do to with radio at all, and, more significant, how to make money out of it, beyond the sale of the boxes. They didn't even know if people would listen to the things. Much like computers in the 1970s, it was uncertain if people would buy these things or put them into their homes. The answer came with a fight.

On July 21, 1921, Jack Dempsey, the heavyweight champion of the world, also known as "the Manassa Mauler," was scheduled to fight Georges Carpentier, the French champion, also known as "the Orchid Man." The bout had gained worldwide attention. Dempsey was a massive fighter, the Mike Tyson of his day. He had amassed 50 wins, with 42 knockouts. Carpentier had 81 wins with 52 knockouts. Although Dempsey was the American and the fight was being held in New Jersey, it was Carpentier who was the hero and Dempsey who was cast as the villain. Carpentier had distinguished himself in the First World War. Dempsey, despite his enormous physical strength, had skipped service in the war and had been registered as 4F, physically unfit for duty. Also, Dempsey had just come off taking the World Heavyweight Championship from Jess Willard in 1919. Willard had towered over Dempsey at 6'7" and weighing in at 245 lbs. Dempsey stood 6'1" and weighed only 187. Yet in the fight Dempsey had pounded Willard unrelentingly and, without exaggeration, within an inch of his life. To this day, the fight is regarded as one of the most savage beatings in boxing history.

Thus it was that mass crowds turned out for the Dempsey-Carpentier match. In Jersey City, all ninety thousand seats were sold out. It would become the first fight in history to exceed a million-dollar gate, and also the first ever to allow women to attend. As it happened, there was also a wireless convention happening in New York City at the time, and as an experiment, it was decided that the fight could also be broadcast on the new radio. The question was, would anyone want to listen?

In 1920, only one in five hundred families in the United States had a radio, so Sarnoff also rigged up more than forty other venues, from Times Square to movie theaters across the country, to receive the live broadcast. The results were astonishing. First, Dempsey won on a knockout, but more significant, the entire country came to a halt to follow the fight as it happened. It was like the *Titanic*, but this time, it was not an accident but a planned and very controllable event that brought in the crowds.

Sarnoff learned a lesson: what people would tune in to. They wanted something exciting. In these early days, Sarnoff was searching for content that would motivate people to buy radios. That was where Sarnoff made his money, in the sale of his radios. Only later would he discover the gold mine that was advertising.

Advertising did not come naturally to radio. As with the Internet in its very early days, radio was looked at as noncommercial and free medium. Then, it all changed. On August 28, 1922, WEAF, one of the first radio stations in New York, advertised a new apartment complex in Jackson Heights, Queens, "conveniently placed near the number 7 subway line." This was the first commercial in broadcast history, and it proved an enormous success. For the complex's $50, the new apartments were swarmed with prospective tenants. Here then, Sarnoff saw the full picture. Radio became an extremely profitable business, not from the selling of radios, which was a one-time event, but from the selling of commercial time, which could go on forever, so long as compelling programming would attract and hold an audience. The trick was in the programming. You had to give the people what they wanted. But what did they want?

It was the answer to that question (which is still asked all the time) that would go on to shape not only the content of radio but also, later, of TV and ultimately the Internet. It was the marriage of the content that was provided (the cheese that attracted the mouse) to the advertising (the trap that snapped shut on his head). But not only advertisers began to realize the enormous power and potential that Sarnoff had unleashed.

HITLER AND ROOSEVELT DISCOVER THE MEDIA

This creation of a device that allowed so low-cost but broad one-way transmission of content would inevitably find its way into political power, as most media would later. Both Franklin Roosevelt and Adolf Hitler would go on to use radio as powerful tools, one a Democrat, the other a dictator.

For Hitler, the radio provided a remarkable device for vastly multiplying his incredible speaking abilities and his verbal powers of persuasion. Before radio, Hitler was limited by the sheer physical demands of setting up mass rallies and speaking before a crowd. Suddenly, with radio, he was able to send his message everywhere, not just in Germany but to the rest of the world. And, of course, radio having taken the course of one-way broadcasting meant that no one was able to respond to him. When the media was cast as a one-way street of information and news, it gave authoritarians and would-be authoritarians incredible control over large populations. It was not for nothing that Vladimir Putin moved early on in his first presidency to take control of all of Russia's TV stations. Once he did that, it was next to impossible for the opposition, when there once was one, to mount a credible campaign against him. De facto, he had a monopoly on the eyes and ears of the entire population of Russia.

For Roosevelt, his fireside chats created a way for him to speak directly to the American people in a way that no president before had ever been able to do. In fact, before Roosevelt, most Americans had never even seen or heard their president in their lifetimes. The office of the presidency, and the man himself were far removed from public view. This was not by design; this was a function of the technology, or rather the lack of it.

Of course, Roosevelt and Hitler had vastly different views, but suddenly, the power to sway the population of an entire nation, as opposed to a room full of people, was made available to politicians. It gave Roosevelt and those who would succeed him the power to bypass, in many ways, the elected representatives in Congress and take their case, for good or for bad, directly to the public on a one-way street.

GROWTH OF THE BROADCASTING BUSINESS

Sarnoff went on to run RCA, the Radio Corporation of America, and later founded NBC, the National Broadcasting Company—both in the radio business. RCA sold the radios; NBC sold the advertising. Radio rapidly grew to become the most powerful medium of communication that the world had ever seen until then, and of course, Sarnoff, as the head of both NBC and RCA, became one of the wealthiest and most powerful men in America if not the world. By the end of the 1920s, radio penetration went from zero to just about every home in America, and for the first time, everyone was immediately connected in a way that people had never been connected before.

By 1929, just prior to the great crash, RCA stock was an astonishing ten thousand times higher than it had been but five years earlier, putting even Apple or Facebook in the minor leagues. The Depression would, curiously, only further drive the popularity of radio. With nearly one-third of the country unemployed, radio was a cheap and popular diversion, something people sorely needed. In that same year of 1929, Sarnoff attended a radio conference where he heard a presentation by a Russian immigrant named Vladimir Zworykin. Zworykin, like Farnsworth, believed that it was theoretically possible to build an electronic television system. If radio had been worth a fortune to Sarnoff, he could see that television would be even bigger than radio.

Sarnoff asked Zworykin how much it would cost to build such a system. Zworykin told Sarnoff that it would cost $100,000, and with that, Sarnoff signed Zworykin, and RCA and NBC were off to television land. Meanwhile, in Los Angeles, underfunded but brilliant Philo Farnsworth had been busy both working on the devices necessary to create electronic television—all of which had to be built by hand by Farnsworth—and filing more than 160 patents for his inventions.

In 1930, Zworykin paid a visit to Farnsworth, and Farnsworth, hoping that RCA would license his invention, pretty much showed Zworykin the farm. Suitably impressed, Zworykin came back to New York and started building what he had seen in Farnsworth's lab without licensing a thing from Farnsworth. In 1931, Sarnoff offered Farnsworth $100,000 for his patents, an offer Farnsworth turned down. Sarnoff decided to proceed without the patents and years of litigation ensued. In the end, David Sarnoff brought television to the public in 1939, launching it at the 1939 World's Fair in New York. "Now we add pictures to the sound," he said.

In the end, Zworykin had been a bit off on his initial estimate of $100,000 to make electronic television a reality. By the time he was finished, Sarnoff had spent more than $50 million, in 1930s dollars no less—nearly half a billion in today's money—to bring television into reality. It was an astonishing investment in what had been, until then, an entirely theoretical concept. Yet his big bet paid off. Within only a few years, television would come to eclipse even radio, and the revenues that television would deliver would dwarf even those that radio had created.

Philo Farnsworth would become a truly tragic figure in American history. He appeared only once on television, the medium he created. On July 3, 1957, he was a mystery guest on the TV game show *I've Got A Secret*. Guests were brought out, and a panel was supposed to, by asking a series of questions, guess what the mysterious person did for a living. Farnsworth stumped the panel. For this, he received $80 and a carton of Winston cigarettes. At the conclusion of the show, host Gary Moore asked Farnsworth what his contribution to television had been. Farnsworth replied, "There had been attempts to devise a television system using mechanical disks and rotating mirrors and vibrating mirrors—all mechanical. My contribution was to take out the moving parts and make the thing entirely electronic, and that was the concept that I had when I was just a freshman in high school in the Spring of 1921 at age fourteen."

As the holder of nearly three hundred patents, Farnsworth was involved in almost constant litigation with Sarnoff and RCA. RCA, of course, had buildings filled with lawyers. Farnsworth and his wife Pem were on their own. Sarnoff made great sport of the ridiculousness that a fourteen-year-old junior high school student could possibly have invented television. But in the end, Farnsworth had his "day in court," so to speak. He was awarded $1 million plus a percentage on every television sold in America. The key piece of evidence that turned the jury? The drawings that Philo Farnsworth had given to Justin Tolman, his high school science teacher. Tolman had kept them all those years and testified in court on Farnsworth's behalf.

Sarnoff went on to build a television media empire that soon eclipsed radio. Farnsworth died in 1971 at the age of sixty-four, an alcoholic, depressed, and virtually unknown. Zworykin would live to the age of ninety-three, dying in 1982. He claimed that he never watched television because it had become so idiotic. "I hate what they have done to my child," he said. "I would never let my children watch it."

Even David Sarnoff probably could not have predicted the

enormous popularity that his new device would generate. At the end of the Second World War there were only ten thousand television sets in the entire country. They were the exotic playthings of the rich; there just wasn't much to watch. By 1950, only 9 percent of US households had TV sets. Yet just five years later, that number had risen to 65 percent, and by 1960, only a decade later, fully 90 percent of American homes had TV sets. According to Robert Gordon, author of *The Rise and Fall of American Growth,* the TV set was the fastest-growing appliance in history, greater than even smartphones and tablet computers.

Suddenly, and almost overnight, Sarnoff and his companions at CBS and ABC, Bill Paley and Leonard Goldenson, had a direct one-way pipeline into pretty much every household in America—effectively a direct pipeline into pretty much every American, as watching TV quickly became *the* national pastime. Despite the Great Depression, NBC, now based entirely on the formula of great programming and sales of commercial time, became one of the most profitable companies in American history, under the leadership of David Sarnoff.

It was only reasonable, then, that when Sarnoff added "the pictures to the sound" in 1939, this model of content that would attract a mass audience, married to the sale of commercial time, would be injected into television. And it was. Sarnoff would turn television into the most profitable money-making machine the world had ever seen. That it had terrible side effects, well, that was not his problem.

In those days, when the TV signal was still analog and was pushed through the air, there was a physical limit to the number of channels that there could be. The size of the wavelength simply ate up the space on the electromagnetic spectrum, so the three networks had the ultimate barrier to entry for any competitor. Essentially it was simply impossible. And, in case anyone would try, they also had the backing of the FCC and the United States government to help enforce their monopoly on the public attention and the public mind.

With one hundred million households in America spending five hours a day watching one of the three networks, any advertiser, by going to television, had immediate and unfettered access to approximately thirty million homes in which to sell its wares.

It was, in a way, unbelievable. It was both an addictive electronic narcotic and a machine to mint money.

There was no way that Sarnoff or Zworykin or Farnsworth could have known what kind of machine they were unleashing on mankind. No one had any experience with this kind of thing. It was unique to the million years of human life that had preceded 1939. Afterward, nothing would ever be the same again. At the end of his life Zworykin said that his greatest contribution to television was the invention of the Off switch.

Alas, once it was turned on, there was no turning it off, and the world of video dominated culture that was to follow. The unique marriage of those flashing lights that we found so very addictive combined with storytelling, which had been the very basis of our survival, proved a surefire, if perhaps ultimately fatal, combination. More than eighty years have now elapsed since David Sarnoff first presented television at the New York World's Fair in 1939. He not only invented a new medium, he invented an electronic drug that would turn not just the nation, but the entire planet, into helpless addicts.

Part II

⊘ ⊘ ⊘

THE ADDICTION

WARNING! ADDICTION TO WATCHING MAY CAUSE DEPRESSION, ANXIETY, BANKRUPTCY, DISSATISFACTION WITH YOUR LIFE, AND, IN SOME CASES, DEATH

7

Living in a Mediated World Where Nothing Is Real

This is your last chance. After this, there is no turning back.
You take the blue pill, the story ends. You wake up in your
bed and believe whatever you want to believe. You take the
red pill, you stay in Wonderland and I'll show you how deep
the rabbit-hole goes.

— *Morpheus in* The Matrix *(1999)*

Without realizing it, David Sarnoff and Vladimir Zworykin had, in
1939, unleashed on the world a terribly addictive and destructive
drug, a kind of electric heroin, only far more deadly. That had not
been their intention.

Radio, Sarnoff's first great project, had been so very successful as
a machine to make money, as a way of keeping people informed and
as a vehicle for mass entertainment, that adding pictures to the sound,
as Sarnoff had explained in 1939 on the launch of the new device,
seemed a reasonable and relatively harmless idea. It wasn't.

No one, after all, spent four or even eight hours a day, every day,
sitting in their living rooms staring into their radios. Who in their
right mind would do that? Radio was the kind of medium that you
could keep on in the background while you were doing other work or
driving your car or cleaning your house.

But television was different. With television, you had to sit down and watch it. Television demanded attention, your complete attention. Radio also did not have the addictive quality of flashing lights. So when television first arrived in American homes in the late 1950s, by the millions, people began to drop all their normal activities and spend more and more time simply sitting in front of their TV sets, staring into them.

This alone might have proven problematic. Staring at a glowing box for hours on end should have been an indication that something fundamental had changed, but it was far worse than simply being addicted to staring. It was also the content that was being spoon-fed directly to millions, hour after hour. While it was true that Sarnoff had, probably without realizing it, tapped into our ancient and until then almost dormant strain of DNA that we all carry that demands that we pay attention to flashing lights whenever we see them, he had then inadvertently added to those flashing lights our deeply inculcated desire for and sensitivity toward being told stories. And it was this killer combination that would prove irresistible.

We were compelled to watch because the need to watch was in our DNA. We were also compelled to pay attention and absorb the lessons we were inadvertently being taught by this new medium because a hundred thousand years or so of using stories to educate and inform was also in our DNA. The results were unavoidable, even if they were unforeseen.

Almost overnight, television viewing went from something that had been nonexistent to our number one activity. This alone is rather extraordinary. In only one generation, this entirely new medium that had not even existed before in any incarnation came, virtually overnight, to dominate almost all of our time and very quickly our lives. The later addition of cable, the Internet, the profusion of screens, and finally screens that we carried with us all the time only increased the supply and made the fixation and the addiction even worse.

It was also ironic, and perhaps tragic, that the rise of the television/video/image culture arrived at exactly the same moment as we began to move from the countryside into cities. It happened by accident, but this was important. Based on the 2010 census, fully 80.7 percent of the population of the United States now lives in urban areas. This is consistent with Europe. The rest of the world is not far behind, but catching up rapidly, with 68 percent of the world's population living in urban areas.

Like staring at screens, this is also an entirely new phenomenon in the way human beings live. According to the 1790 census in the US, for example, fully 95 percent of the population lived in rural areas. That, of course, was entirely congruent with the way human beings had lived their lives since they first appeared in East Africa. It is only in the past very few years that this has changed. Urban living, as with watching the world on screens, is an entirely new phenomenon for our species.

For some two million years (based on fossil finds in East Africa), human beings had lived in direct and immediate contact with their environment and the real world around them. When they walked, they felt the wet earth beneath their feet; they smelled the plants; they grew sensitive to the sounds of the wildlife, to the call of birds; they felt the sun on their faces, the tactile sensations of the forest or the jungle or the desert. They had a direct and personal contact with the food that they grew and ate. As the book of Genesis says, "In the sweat of thy face shalt thou eat bread, till thou return unto the ground; for out of it wast thou taken: for dust thou art, and unto dust shalt thou return." They lived real lives.

In this way, life in the Paleolithic was little different from that in biblical times to Greek or Roman life to medieval life, to even Elizabethan life. People lived in daily and extremely intimate contact with the real world. When the Bible says, "in the sweat of thy face" it was literal. Almost everyone worked with their hands, and most people with the soil or with animals. Today, most people work in a cubicle or an office, spending their days staring into a screen, trapped in a hermitically sealed building. When they go home, their lives are little different: different room, different screen, but fundamentally the same, and fundamentally a million miles from the way their ancestors for one hundred thousand generations lived.

THE BIG CHANGE

The rise of the media culture brought about a massive change. And massive changes are traumatic.

In 1967, psychiatrists Thomas Holmes and Richard Rahe created what is popularly known as the Social Readjustment Rating Scale. They asked people how stressful they found forty-three different events, and then rated them. Based on their findings, the most stressful events in life are things like death of a spouse, a marital separation, moving houses, and so on. They did not rate being ripped from your natural environment for the past million years and finding yourself in a totally new and alien world. Now *that* is stressful. And that is exactly what we have all undergone just in the past sixty years. Is it any wonder that you feel that something is missing in your life, but you can't quite put your finger on what it is?

If you feel that way, you are not alone. Study after study shows that a vast majority of the population today is in some way fundamentally unhappy if not yet mentally unhealthy. A recent study by the Center for Clinical Interventions found that 70 percent of girls "feel they are not good enough." A 2017 Gallup poll found that 85 percent of Americans hate their jobs. Approximately one in eight Americans are taking some form of antidepressant drug. Clearly something is not right.

It is because we were not meant to live like this. We are not, by dint of evolution, programmed to live like this. We are the children of one hundred thousand generations of an evolutionary process that had exactly the same contact with reality, lived the same experiences. It was this kind of world that made us who we are now. This is deep inside of all of us. And then, in the blink of an eye, our world was changed, or rather we changed it.

The urbanization of human existence, both in terms of home and work, meant that we would now spend almost all our time living and working in tiny boxes, one on top of the other, reaching into the sky. Our day-to-day environments are no longer the forest, the fields, the open air, the sun. Instead, they are concrete and glass and carpeting; air conditioning and fluorescent lights. Often, even our windows no longer open. Between our apartments and our offices, we can spend

days if not weeks on end never breathing real air. We travel in completely enclosed and totally mediated environments, whether they are
cars or airplanes. Our homes are sealed off from the outside world.
Our food is manufactured. And, our only contact with the outside
world is through what we see on our screens. When we do walk in the
streets we cut off contact with the real world with ear buds. More and
more, the media and the mediated world are our only knowledge of
what a forest is like, what the ocean is like, what a desert is like, what
animals are like, what insects are like, and more often than not, what
other people are like—particularly people who are not like us. For us,
those images are real. We see and we hear that which has been carefully produced and packaged for us by someone else. And because our
contact with the outside world has been mediated, we no longer have
to process information. We are simply passive receivers of the way
someone else wishes us, with our now truncated senses, to have limited contact with a created reality. The Pirates of the Caribbean ride
at Disney World and the *Pirates of the Caribbean* movie have about as
much relationship to the actual pirates of the Caribbean as an orange
soda has to an actual orange. Yet ask someone to describe a pirate to
you, and that person will invariably describe Johnny Depp. (I have
done this experiment more than one hundred times.) To them, that
is a pirate.

Our new way or living, that way of perceiving and receiving the
world, that never-ending immersion into a two-dimensional landscape
of carefully packaged and prescreened sounds and pictures is so far
divorced from our prior two million years of experience that it is as
though we have all been plucked from the lives we were designed and
destined to live and deposited on some alien world with which we have,
in fact, nothing in common. Is it any wonder that you feel anxious or
depressed or somehow, in a way you can't quite explain, disconnected
from real life? It's not neurotic. It is, in fact, kind of a healthy response
to the world in which you are living. You don't need Zoloft or Xanax
or Klonopin to suppress your feelings. Your feelings of angst or dread
or depression may be the most honest feelings you have.

There is a common complaint now that we are raising a generation that is singularly unprepared for the "real world." These are

called helicopter parents, because they hover over their children's every activity to make sure they are safe. In a highly mediated world such as ours, these anxieties about their safety and their reaction to continually keep a watchful eye are not neurotic. They are, in fact, an almost natural reaction to their perception to a world out of kilter. They spend their lives living in a highly mediated world of either sealed homes and offices or two-dimensional presentations of what the rest of the world is actually like (to which, of course, it bears almost no relationship).

Their child, sitting safely at home, watching a video on YouTube or killing hundreds of people on a video game, inhabits a remarkably safe world, and one that is largely congruent to the world the parents also inhabit. However, let the child put one foot into an unmediated and clearly unguarded and unprotected world and, of course, who knows what could happen.

Well, of course the inhabitant of the mediated two-dimensional world knows all too well. Is not the news filled with horrific stories of pedophiles on the loose? Of kidnappers? Of God only knows what kind of people who are wandering the streets at random and liable to kill, rape, maim. or otherwise injure your child, if only they are given half a chance? So the terror of the parent is not neurotic. It is a very reasonable reaction to the world as they both live it and perceive it.

Have you ever seen the movie *Jaws*? More than a billion people around the world have seen the movie, either in theaters, on TV, or online. It's a great story, and Steven Spielberg is a master storyteller. And because we are innately wired to understand on some level that stories are told to educate us, we all know that sharks are out to eat us, if they can, given half a chance. As Dr. Hooper, played by Richard Dreyfuss said, "Out there is a perfect engine, an eating machine that is a miracle of evolution. It swims and eats and makes little baby sharks, that's all."

If stories are there to educate us, which is what they have always been about, then in this magnificent scene, Spielberg and Dreyfuss give us a first-class education in the life of a shark. That education is, of course, compounded greatly by actually seeing people getting eaten by sharks in the movie. Who needs to go to Woods Hole to get a PhD in ichthyology? The movie is so much better a teacher.

And the result? Well, we have, now a well-learned and deep-seated fear of sharks and of swimming in the ocean. God only knows what could happen. Well, of course, you do know what could happen. You or your kid could get eaten. And, in a perfect meshing of media to media, every time some poor bastard does get bitten or eaten by a shark, it makes front-page headlines (or fills the blogosphere) for a day or two.

According to the ISAF (International Shark Attack File), your odds of getting attacked by a shark are 1:11.5 million. Just to put that in perspective, your odds of getting killed in a car crash are 1:84. Terrified to get into the ocean? You should really be terrified of getting into a Ford. Of course, we don't get a lot of movies about people getting dismembered in car crashes, even though this happens a lot more often than a shark biting off your arm. The reason, of course, is that the Ford Motor Company buys a lot of ads on TV; sharks, not so much.

So our highly mediated world presents us with a fascinating dichotomy. Clearly, the "real world," as perceived through the media, is incredibly dangerous. But, if we only continue to experience it just through screens, we are all safe. No one died, so far as I know, from watching *Jaws*.

The problem with living in a mediated world, and experiencing most things through a two-dimensional screen using only two of your senses, is that you are innately and very much by design distanced from any contact with reality. As a result (as in the case of *Jaws*, for example), you have no personal experience upon which to make judgments. Your contact with the real world has been cut off, not just with sharks and pirates, but with everything. After two million years of living in the real world, don't you think this would be anxiety-inducing?

Many of us are suffering, in a sense, from a kind of post-traumatic stress disorder. You understand that soldiers get PTSD when they are pulled from their homes and their families and the lives in which they are comfortable and suddenly placed into war zones in which they are expected to kill people or be killed. *There is something fundamentally wrong here*, their brain is telling them, even if they don't want to believe it or understand it. So it is with us.

There are many upshots to this kind of existence, none of them good. Perhaps the primary point here is that when you have come to mistake mediated experiences for real life, you will soon find any contact with real life both confusing and terrifying, and in the end, you will want to either stay away from it or make sure that it is made "safe" before you get there.

Since the end of the Second World War in 1945, we have all collectively lived through the greatest economic expansion in human history. We have also lived through more than seventy years of relative peace. The invention of nuclear weapons in 1945 meant that for the first time, human beings had the power to kill on a scale that was previously completely unknown. For all of our history, warfare had been conducted, more or less, on a relatively personal level, and with great ease. Nations and their armies went to war at the drop of a hat. They spent their time preparing for it, taxing for it, rehearsing for it, and glorying when it arrived, which was with a great deal of regularity. The Thirty Years War, the Hundred Years War—we had never-ending glorious battles and wars. But the advent of nuclear weapons meant that whole cities, hundreds of thousands, and then millions of people, could be wiped out in an instant, at the touch of a button. This was something entirely new. The fact that not only were there nuclear weapons but also the two great adversarial powers had them in massive numbers, meant that no one would ever use them. Not only was nuclear war off the table, so too was war between any major powers. It was now too unthinkable.

For almost all of human history, going to war had been a regular part of life. Germany went to war with France; France went to war with Britain; Britain went to war with Spain, Greece went to war with Athens, Macedonia went to war with Persia. But after 1945, no great power ever went to war with another great power again. It was clear no one could win. And so we entered a fascinating era of both peace, with a few limited skirmishes that bore nothing comparable to the First or Second World Wars, and of massive economic expansion.

The postwar economic boom is called by economists the Golden Age of Capitalism, and with good reason. All over the world people grew rich—and safe. Average people began to acquire things and live lives that would have been unthinkable for the totality of the rest of human history. The average worker in the United States and Europe lived better, longer, and a more luxurious life than most medieval kings. It was an astonishing time. With things so well, you would think that people would also have been extraordinarily happy. The irony was that concurrent with that vast and global expansion also came an era of vast unhappiness, depression, anger, and frustration.

In 1947, W. H. Auden published *The Age of Anxiety*, but he was early. In 1947, the age of anxiety had not even yet begun. If we were to chart the world and what we might call the state of unhappiness with life, it is shocking to see that between 1950 and the present pretty much every graph that is supposed to measure these things takes a hockey-stick upward jolt. Why is that? Why is what should have been the best of times for everyone actually a time of vastly increased stress, anxiety, depression, drug use and abuse, divorce, imprisonment and so much more? What could possibly explain it all? Why would people who are living so well, really, for the first time in all of human history, be so fundamentally unhappy?

The Nielsen Company, which does the television ratings, charted the average hours of TV watched daily over the years. It followed this hockey-stick upward spike.

Clearly, prior to the 1940s, the number of hours watched would be zero, as the medium simply did not exist. But as we enter the post-war 1950s, and television sets become more and more ubiquitous in our lives, the number of hours of watching increases rapidly. By 2010, the peak watching year, the average American is spending an astonishing nine hours a day watching TV. From 2010 until now, TV hours have tapered off somewhat, but only because that time has been replaced by watching videos online.

Now, if you compare hours of TV watched to opioid abuse in America during the same time frame, you'll find much research following the same upward pattern.[10]

Studies show that rates of depression for Americans have risen

---

10 Leslie Dye, "Opioid Epidemic Resource Center," Elsevier website, updated January 29, 2020, https://www.elsevier.com/connect/opioid-epidemic-resource-center.

dramatically in the past fifty years.[11] The explosive increase in depression (and anxiety) almost exactly mirrors the explosive rise in number of hours a day people spend watching TV and videos. Are you starting to see a correlation?

We can also take a look at obesity rates in the same period. According to the CDC (Centers for Disease Control and Prevention), obesity in America has skyrocketed by 185 percent over the past fifty years, almost exactly tracking the rise in screen watching in America.

The New York Fed Consumer Credit Panel and the Federal Reserve Board have tracked personal debt over the past fifty years. That research tracks pretty much along the same lines of TV- and video-watching. Personal and consumer debt has also skyrocketed since the arrival of your own personal home media screen systems.

One more upward-pattern phenomenon comes from the Bureau of Justice Statistics. From 1880 to the present day, there has been a hickey-stick-patterned increase in US prison populations—once again, the vast spike in incarceration that matches the arrival of the move from contact with the real world to perceiving the world through media.

I could go on and on with these, but I think you get the point.

In the subsequent chapters, I am going to deal with each of these phenomena, as well as others, to explain why this transformation in our society, and the concurrent sense that "something is wrong," may be laid at the feet of our ubiquitous and all-encompassing world of media.

---

11 Chris Iliades, "Stats and Facts about Depression in America," Everyday Health website, January 23, 2013, https://www.everydayhealth.com/hs/major-depression/depression-statistics/.

8

Junk Food, Junk Media

"Tell me what you eat, and I will tell you what you are."
—*Jean Anthelme Brillat-Savarin*

If you are going to spend a good deal of your life immersed in the media, which it seems you are, then here is something you should know about the media: The media does not exist for you. You exist for it.

In the world of the media, you are not the client; you are not the audience. The TV shows, movies, videos, GIFs, boomerangs, and any other moving images that occupy your time are not there to entertain or inform you. They are there to capture you. You and your data are the product being captured and sold to advertisers. You may think of this as digital slavery, which is what it is.

Media companies, whether Facebook or Instagram or CBS or Disney, don't get paid by you; they get paid by advertisers. The media companies earn their vast incomes through selling advertising and data. Advertisers pay those media companies to get access to your eyeballs and to your data, and thus you. The more eyeballs, the more viewers, the more they pay. The job of the media company is to attract you to their content and hold you there so that the advertisers can slip in their own content, which is specifically designed to convince you to buy their product.

Do you get the concept here? The media company builds the mousetrap. The content is the cheese. You are the mouse.

If you saw the movie *All the President's Men* about the Watergate scandal and the *Washington Post*, the catchphrase for the movie and the investigative reporting done by Woodward and Bernstein was "Follow the money." This works pretty well, whether you are talking about presidential corruption or the corruption of a culture.

Let's look at drugs, another very profitable addiction-based industry. It's a nice model for the media business. If you want to understand the architecture of the drug problem in America, you can follow the money. The people who make the money here are the drug suppliers in Colombia. The recently arrested El Chapo Guzman was worth an estimated $14 billion, but that pales in comparison to Pablo Escobar, who, in 1989, was declared by *Forbes* as the seventh richest man in the world with a net worth of $25 billion, or $52 billion in 2019 dollars.

If you think about it, the drug business is pretty impressive. The whole massive structure of the illegal drug industry—heroin, cocaine, meth, and so on; the billionaire drug lords in Colombia; the millionaire kingpin dealers in New York or Los Angeles; the lavish lifestyles; the yachts, the private jets, submarines, and speedboats to slip the stuff in past customs; the massive armed compounds; the private armies—all of it is 100 percent financed by the individual junkie stealing money from his mother's purse. That's the bottom of the drug chain, but that's where all of the money comes from. There are just a lot of small contributors—a whole lot of $20 bills being slipped out of purses.

Addictive businesses can be very profitable. The cigarette industry is worth billions, and that all flows from some poor schmuck lighting up his Marlboros and forking out a few bucks for a pack over and over and over again—just like the drug dealers, lots of small contributors.

But when it comes to profitable addictive businesses, cigarettes and illicit drugs cannot hold a candle to the profits made from our media addiction. The global media business alone is worth some $1.9 trillion annually. That makes it bigger than global oil, which you know is pretty big, and that does not even begin to factor in the value of the data being sold, also a byproduct of your addiction, from videos on YouTube, or Facebook or Instagram or any other platform. Overall, our video addiction is a huge and powerful global industry.

Like the junkie on the street, the vast wealth of the global media empire comes down to you, the individual user, stealing money out of your mother's purse for yet another fix. You are constantly doling out dollars in small increments, relatively, to the media companies, but not on the street. You are doing it in the comfort of your home, on your couch, online, on Amazon or in Walmart.

The reason that advertisers are willing to put billions and billions into buying media ads is because those ads drive us to purchase stuff. If we didn't buy stuff in vast quantities, then companies would not spend billions on ads. And if there were no ads, there would be no content and no networks and no cable. Did you know that a little-watched cable channel like DIY takes in more than $100 million a year in ads? And that is for a cable channel that you probably never watch. Imagine what the networks and channels that you *do* watch take in in ad revenue.

Clearly, the ads work, and it is the ads that endlessly compel us to buy more and more stuff that we don't need. Increasingly, the more blatant ads are being replaced by product placement or influencers. It matters not. The object is to grab you by the collar, drag you into a room, and convince you to spend your hard-earned money on a bunch of stuff you don't really want or need. You just think you do.

Jerry Mander (*Four Arguments For The Elimination of Television*) notes that the sole purpose of advertising is to convince you to buy things you neither want nor actually need:

Sales, by definition, is the process of convincing someone to purchase what they don't need. Advertising tries to convince someone that the solution to a problem or the fulfillment of a desire can only be achieved through the purchase of a product. If we take the word *need* to mean something basic to human survival—food, shelter, clothing—or basic to human contentment—peace, love, safety, companionship, intimacy, a sense of fulfillment—these will be sought and found by people whether or not there is advertising.

When you look at it objectively, it makes drug dealers look a lot more attractive.

WHY THIS WORKS

We are massive, unstoppable consumers. We are Olympic-class shoppers. Like locusts, we consume everything in our path.

Not so very long ago, a family might have very few treasured possessions: a few pieces of furniture, hand-made and carefully passed down from generation to generation; a few articles of clothing; one pair of shoes, perhaps two for Sunday best; a watch; some cutlery and utensils; and that was about it. Material things were hard to come by, and if people were lucky enough to get them, they held on to them, took care of them, and passed them on to their progeny, expecting they would do the same. If items broke, they fixed them. Possessions were, like pre-printing-press stories, few, respected, and cherished.

In the eighteenth century, for example, the average home in America or Europe measured just 450 square feet. Today, the average home measures 2,687 square feet, while, ironically, families have gotten much smaller. But the houses aren't there to hold our families; they are there to hold our stuff.

You may have heard of the Pareto Principle: the 80/20 rule. That is, you wear 20 percent of your clothing 80 percent of the time. This is probably true. If you open your closet, you will no doubt find a massive excess of things that you bought but never wear. This could account for the sudden popularity of Marie Kondo and her *Life-Changing Magic of Tidying Up*, which encourages tossing out stuff you bought but never wear or, in her words, doesn't give you "joy."

As it happens, the world now consumes over eighty billion pieces of clothing each year, up 400 percent from just two decades ago. The average American throws away some sixty pounds of clothing each year—and that was before Marie Kondo. Clearly, we are buying far more clothing that we could ever need or want or even wear. So why do we do it? We do it because the media is adept at convincing us that buying new clothing (or shoes or cars or phones or TV sets or furniture) is the pathway to a happiness that seems perpetually unattainable.

This is what is called "retail therapy." It doesn't work, at least not for you. For those who make and sell clothing, shoes, cosmetics, phones, cars and everything else, it works great. As a consequence of this "therapy," household debt in the United States reached a staggering $13.5 trillion last year. Credit-card debt alone surpassed $1 trillion, with the average American household owing $16,883 on credit cards. These debts most likely are never going to be paid off, simply perpetually serviced at 19.24 percent, while store cards average 24.75 percent interest.

In 2018, the Federal Reserve Board reported that nearly half of all Americans could not come up with an additional $400 if they needed it for an emergency. In an article in the *Atlantic* in 2016, successful author and film critic Neal Gabler admitted that he was in that 47 percent who could not come up with the $400. And he is hardly alone. Gabler notes that if an average American family making $50,000 a year were to lose their jobs, they would have enough money to last them six days. So many people are living on a knife's edge, just barely managing paycheck to paycheck. It should not be this way, but it is.

Our greatest consumption, however, is not clothing; it is food. Eating is by far our favorite thing to do, after watching. Eating beats shopping, but not by much. And often eating and watching go together. A tub of Ben & Jerry's, a spoon, and a movie on Netflix— what could be a better way to spend a rainy afternoon, or a lonely evening, or an evening with family, for that matter? Maybe a bag of Doritos would be better. Or maybe order in a pizza? All of these things seem appealing, but no one ever says, "Hey, let's binge watch *Game of Thrones* and have a nice bowl of broccoli."

But if you are going to eat all the time, and particularly bad things, you are going to pay a price. Perpetual shopping keeps you perpetually living in debt; perpetual eating makes you fat. That's just how it all works. All of this is driven by the media. You want to blame someone that you are overweight; don't blame your mother, blame your TV set or your phone.

Prior to 1950, when TV sets first began to appear in American households, barely 1 percent of the US population was classified as obese. By 2016, however, 39 percent of US adults were classified as obese. That is an astonishing change, and the skyrocketing rate of obesity in the US parallels the arrival of television and its equally sky-rocketing explosion of TV- and video-watching. There is a reason for this correlation.

You might think that the correlation here is due to spending more time on the couch watching things, as opposed to going outside and getting exercise. That is, in part, true. The arrival of a culture of watching did lead to a decrease in physical exercise of almost any kind. But the correlation I want to make is, I think, far more interesting and far more relevant.

For almost all of our history on this planet, food was primarily about survival. Could you get enough food to eat so that you could simply live? This was the primary preoccupation of just about every human being since we first learned to stand erect. Food was hard to come by. Survival hinged on getting enough food every day.

As a result, for almost all of our history, starvation was never too far away. Famine was a constant companion of life and a very real threat. A bad harvest or two could kill a person and the family. It was pretty simple. There is a reason that one of the four horsemen of the Apocalypse is famine. There is a reason that when God expels Adam and Eve from the Garden of Eden, he curses them by saying, "You shall eat your bread by the sweat of your brow." That is the reason that pretty much all the ten plagues that God rains on the Egyptians in the book of Exodus are actually related to food: turning the Nile to blood, frogs in the bread ovens, pestilence of the livestock, and so on. It's almost all about food and famine.[12] By the time we get to plagues nine and ten, darkness and death of the first born, it is pretty much a given that after the other eight, there is not much left to eat in Egypt anyway.

The core of the story—and it is, of course, a story with a message that is still repeated to this day—is that unless you, pharaoh, release the Israelites, your people will starve to death. The entire political and religious structure of ancient Egypt was based upon the pharaoh's abil-ity to control the periodic flooding of the Nile, that which watered

12 You'll note how the power of the story is greatly enhanced by tying it to the no-tion of food and starvation, an omnipresent fear in most of human history.

the crops, and thus that which provided the food. Earlier in the book of Exodus, of course, Joseph is taken from his prison cell and brought before the pharaoh to interpret the pharaoh's dreams. He dreams of seven fat cows followed by seven emaciated cows; seven withered ears of corn that devour seven fat ears of corn. Joseph interprets the dream to predict seven years of good harvest followed by seven years of famine. The pharaoh, acting on Joseph's interpretation, builds warehouses to store the grain in the good years, thus saving his people and his rule.

For almost all of our history, not just in biblical times, getting enough to eat was hard work, and often perilous. Survival was always uncertain, and starvation was almost always at the door. In 1315–1317, for example, Europe suffered what is called the Great Famine. Seven-and-a-half million people simply starved to death. That was out of a total population of 78 million. Today, it would be the equivalent of 80 million people in Europe dying of starvation in one year. In 1601–1603, a massive famine swept Russia. About two million people died of starvation. In France, in 1693–1694, one-and-a-half million people died of starvation. In 1783–1784, more than eleven million died in India during the Chalisa Famine. Between 1850 and 1873, an astonishing twenty million people died of starvation. In 1932–1933, under the rule of Stalin and his move to collectivization, at least 3.9 million Ukrainians starved to death. As recently as 1984–1985, about 1 million Ethiopians died of starvation.

Starvation has always been with us throughout our history. Thus, we have a natural and instinctive urge to eat as much as we can, when we can, because you never know when the next meal is coming. This drive is deep in our DNA. For pretty much all of our history, there was a direct correlation between eating and surviving to reproduce. Those who could find enough food to eat lived. That was a trait that they passed on to their children.

There was a time when almost all of us grew all of the food that we ate. That was where our food came from—from the sweat of our brows, from our own hands. As recently as 1870, almost half the country worked in agriculture. We knew food. We knew how to grow it, how to harvest it, how to prepare it. We had an organic relationship with what we ate. Today, fewer than 2 percent of the population works in agriculture. We are distanced from what we eat. Instead of growing it ourselves, our food today is provided for us by massive food companies. They manufacture, process, and deliver most of what we eat now. We are in 100 percent in their hands.

And who are these companies, and what is the food that they make? This is worth a moment's consideration, because it says a great deal about who we are. These figures are from 2016:

1. Nestle, with $90.2 billion in revenue: DiGiorno, Hot Pockets, Butterfingers, and Kit Kats.
2. Pepsico, $62.8 billion: Pepsi and other sodas, Cheetos, and Tropicana.
3. Unilever, $48.3 billion: Magnum Ice Cream, Hellmann's Mayonnaise, and Lipton Iced Tea.
4. Coca-Cola, $41.9 billion: Coca Cola, Diet Coke, Cherry Coke, Fanta, and Sprite, which includes four of the world's top-five selling soft drinks.
5. Mars, $35 billion: M&Ms, Starburst, Orbit gum, and Uncle Ben's rice.
6. Mondelez, $25.9 billion: Oreo cookies, Trident gum, and Sour Patch Kids.
7. Dannon, $23.7 billion: Activia, Yocrunch, and Dannon Yogurt.
8. General Mills, $16.6 billion: Cheerios, Yoplait, Hamburger Helper, and Haagen-Dazs ice cream.
9. Associated British Foods, $16.8 billion: Dorset Cereals and Twinings Tea.
10. Kellogg's, $13 billion: Corn Flakes, Froot Loops, Frosted Flakes, Eggo, Pringles, and Cheez-Its.

As you can see from the list above, the "foods" that these companies are best known for are far from healthy. What they are, however, are best-selling products. And the reason they are such big sellers and such big profit centers is that they appeal, in a very direct way, to our fundamental cravings for things like sweetness, crunchiness, a certain silky feel on the tongue, and so forth.

These companies are not so much interested in food, per se, but rather first in making a nice profit. They make their profit by selling us as much of their stuff as they can. And how do they manage that? They have massive buildings filled with scientists who are continually fine-tuning the "food" that they are selling to us. They are searching for the formula that makes it both attractive and addictive.[13] And they are very good at this.

If you remember the old Lay's Potato Chips ad, "Betcha can't eat just one" meant that surely you could not stop at just one. And, as anyone who has cracked open a bag of potato chips can attest, once you eat one you are pretty much committed to the whole bag. Of course, this is not true just for potato chips but also pretty much for everything you stick in your mouth, from KFC hot wings to Big Macs to Ben and Jerry's Cookie Dough, to pizza to microwave popcorn to Cocoa Puffs. It won't kill you, at least not immediately, but it will keep you coming back for more, and that is the whole idea.

Now, we eat this stuff by the truckload because it cleverly taps into a very old and deeply buried bit of our DNA that is a remnant of our million years on the planet or so. As people died of starvation, those who survived were driven to eat. And what were we all driven for? Well, the really prized find was something sugary, like honey—lots of energy packed into a small space. So what do food companies pack their industrially produced and processed food with? Sugar—tons of it. And why? Because it taps into that very deep DNA that said, "You have to eat this to survive."

Only in the past fifty years or so has eating sugar been considered a death sentence. It's an idea antithetical to a million years of learned survival. In a wonderful book *Sweetness and Power: The Place of Sugar in Modern History* (Viking, 1985), historian Sidney Mintz outlines how our addiction to sugar basically drove most of Western history. Today, the average American consumes his or her body weight in sugar.

The other factor key to survival for millions of years was fat, if people could find it. As it turns out, today sugar and fat are the staples of the American diet. For most of our history as a species we

13 I am deeply indebted to Mark Bittman, noted food author, who explained all of this to me.

could not get enough. Today, we can get as much as we want. Take a walk through any supermarket in America and look at what is on the shelves passing as food: Honey Combs cereal, Sugar Smacks, Sugar Pops, Oreo Double Stuf—fat and sugar.

And what are you gonna do with all that stuff? Why, you're going to eat it! The reason we are so quick to scarf down all these chemical concoctions is two-fold: First, they clearly tap into our very ancient desire to survive, but second, and far more significant, is that they are acceptable to us as food because we are all now so distanced from what real food looks like and tastes like.

No actual potato farmer would eat Pringles; they wouldn't recognize it as food. No dairy farmer would consume a container of Cool Whip. When most of us worked in the agricultural world, when we grew our own food and ate what we grew, it is unlikely that you could have passed off a Twinkie as something edible. Twinkies and Hostess Cupcakes and Snickers Bars and Chicken McNuggets all came of age after the Second World War when only 2 percent of us actually had any real hands-on experience in food.

In a recent study by The US Department of Agriculture, an astonishing 48 percent of Americans did not know where chocolate milk comes from. Seven percent thought that it came from brown cows. Seventy-two percent of children did not know that cheese comes from milk.[14] We are separated from the sources of our food, and we have placed the production and distribution of our food in the hands of a very few giant corporations.

AS GOES FOOD, SO TOO GOES THE MEDIA

The French philosopher Jean Anthelme Brillat-Savarin wrote, "You tell me what you eat, and I will tell you who you are." In America, we might more properly say, "You tell me what you watch, and I will tell you who you are. You tell me your favorite movies, TV shows, or videos on YouTube, and I will have a pretty good sense of who you

---

14 Laura Northrup, "Dairy Industry Says 48% of Americans Don't Know Where Chocolate Milk Comes From," Consumer Reports website, updated June 15, 2017, https://www.consumerreports.org/consumerist/10281160/.

are." This, after all, is the basis for the famous algorithms that Netflix, Facebook, YouTube, Amazon and others have developed.

Netflix spent a fortune creating an algorithm that could predict what movies or TV shows you would like based on what your personal prior viewing habits had been. If you shop on Amazon, you will note that as make your purchases, Amazon will suggest other purchases you might like to make. This is less analogous to the junk that supermarkets place right before the cash register and far more analytical, based upon your prior buying habits. As it turns out, one third of all purchases made on Amazon are a derivative of these algorithm-based offerings.

Today, as with our food, our media is manufactured for us by giant faceless corporations whose only motivation is profit. Their concern is not about your being well informed or educated, no more than the food companies are motivated by a desire to make sure you are eating healthy and nutritious meals. The food companies don't care how healthy you are, so long as you buy their stuff, and the media companies don't care how well informed you are, so long as you watch their stuff. They both want to fill you up with empty calories, carefully engineered to make you crave more and more of them.

The people who eat a tub of Ben and Jerry's ice cream every night know what they are doing, more or less. Maybe you feel bad after you have consumed a half-gallon of Ben and Jerry's Cookie Dough ice cream in one sitting. While you are eating it, you simply cannot stop. The combination of the texture, the creaminess, the slickness on your tongue, the sweetness, the pleasure of finding those lumps of cookie dough—it all adds up to an addiction, until you find yourself scraping the bottom of the container with a spoon to make sure you have extracted every last bit of that creamy goodness. Maybe after it's all gone you run your finger around the container just to make sure you have not missed any. That's an addiction. But once it is over, you sit back and say, "I cannot believe I just did that." But you did.

The same holds true once you start to binge-watch TV or videos. Binge watching is an entirely new phenomenon, brought to you by the same people who create the content. It's like binge eating. You simply cannot stop. But then, after three or five or eight hours of

House of Cards, you sit back and say, "I cannot believe I just did that." But you did. And it is as bad for you as the Ben and Jerry's—maybe worse.

Junk food leads to an obese nation. According to a study by the Harvard Medical School, 50 percent of children today will be obese by the time they reach the age of thirty-five.[15] That's thanks to industrialized food. But junk content leads to a nation of fatheads and unhealthy minds. And that is far more dangerous. And they don't die from bad information. Instead, they go on to make bad choices, based on that bad information. This is particularly true with our news and information programs, our cable news channels, our online news feeds.

Yet the architecture of a profit-driven news and information business almost forces the news companies to create content that will attract and hold the largest audiences, no matter what the consequences. This is dangerous.

We have, in the past few years, come to finally recognize the dangers that industrialized food and its natural consequence of obesity engender. The world is replete with diet plans. The diet business is a $20 billion a year industry with one hundred million Americans dieting. We at least know there is a problem.

However, there is, as yet, no Brain Watchers to correct the damage that industrialized content has caused. What we have created is what we might call the entertainment-news-advertising complex. This was done without our ever realizing it was happening, but now it has an enormous impact on our lives.

It is one thing to binge-watch *Game of Thrones* or *Peaky Blinders*. They may be addictive, and you innately understand that this is fiction. But what happens when these addictive qualities are applied to the news?

---

15 Claire McCarthy, "More Than Half of Today's Children Will Be Obese Adults," *Harvard Health* (blog), December 5, 2017. https://www.health.harvard.edu/blog/more-than-half-of-todays-children-will-be-obese-adults-2017120512879.

9

Fake News

"This instrument can teach, it can illuminate; yes, and it can
even inspire. But it can do so only to the extent that humans
are determined to use it to those ends. Otherwise it's nothing
but wires and lights in a box."

— *Edward R. Murrow, 1958*

In 490 BCE, the Persian Empire invaded Greece. The Persians had
a massive army of more than twenty thousand armed men and 600
ships. The only defense Greece had was the Athenian army of five
thousand. Things did not look good for Greece. The two armies met
at a plain outside of Athens called Marathon.

In order to get help, the Athenians dispatched a messenger named
Pheidippides to run to Sparta, some 150 miles away, to ask if they
would help. The Spartans said they would like to help, but they only
fought during a full moon, which was still six days away. Could the
Athenians wait?

Pheidippides then ran another 150 miles back to Marathon to
deliver the bad news. As it turned out, in one of the most miraculous battles in history, the Athenians actually defeated the Persians
and sent them back home. Delighted at their victory, the Athenians
then tasked Pheidippides to run another twenty-six miles to Athens

to deliver the good news. Upon arriving in Athens, Pheidippides announced to the city, *"Nike! Nike! Nenikekiam!"* ("Victory! Victory! Rejoice, we conquer!") and promptly died. From this comes our modern-day marathon run, the same twenty-six miles.

Nearly two thousand years later, Napoleon, the recently deposed former Emperor of France, escaped from Elba. Most of Europe had been at war with Napoleon since he first appeared on the scene shortly following the French Revolution in 1789. In a relatively short time, Napoleon had gone on to conquer most of Europe in a never-ending series of wars and battles.

No one had fought Napoleon longer or harder that the English, and when Napoleon was at last defeated and exiled to Elba in 1815, Britain rejoiced. Now, news of his escape sent the British stock market into a frenzy. Napoleon had not only escaped from Elba but also had landed on the south coast of France, gathered up an army, and was making his way back to Paris to reclaim his crown and throne. Thousands were joining him and following him. If he was successful, it would mean years of war between France and England. In order to stop Napoleon, the British had dispatched the Duke of Wellington with an army. They would meet at Waterloo, in Belgium, where the future of Europe would be decided.

Information did not travel fast in 1815—in fact, not much faster than it had at the time of Marathon—so people had to wait to hear what the fate of Europe and their own would be. Meanwhile, the London Stock market gyrated wildly on every rumor.

One person who was particularly concerned about the outcome at Waterloo was Nathan Rothschild, head of the British branch of the family bank that bore his name. The Rothschilds had invested heavily in British gilts, or bonds, and a British defeat would be crippling.

Rothschild had recently been approached by an eccentric young inventor who had an idea. Rothschild agreed to see him and hear him out. He arrived with a pigeon in a cage. He would take the pigeon, he explained, and accompany Wellington to Waterloo. Once the battle was over, he would release the pigeon, it would fly directly from Waterloo back to Rothschild's office, and, in a band wrapped around his leg, carry the result of the battle. Rothschild would get the news three days ahead of everyone else.

Rothschild, of course, immediately agreed to invest in this radical new technology for transmitting news. The battle concluded, the pigeon was released, it flew to Rothschild as promised, and, three days ahead of everyone else, Rothschild got the news.

Once he read what was on the band, he headed down to the floor of the London Stock Exchange and quickly announced that Wellington had been defeated and Napoleon was triumphant. On hearing this, or so the legend goes, stock prices immediately collapsed. Rothschild then swept in to buy up as many shares as possible. Three days later, when the first ships from Belgium arrived in England, they carried the news of Napoleon's defeat. Stock prices rocketed. Rothschild, of course, made a fortune.

True or myth, the story is the first indication of the ability to monetize news. News has always been with us. But when you marry news to the ability to make money, you turn information in to a commodity that can have enormous value. This monetization of news is entirely a function of technology. In Rothschild's case, that technology was the pigeon.

THE GUTENBERG EFFECT

In 1450, Gutenberg printed his first Bible. The day after he printed his first Bible, he could have printed the *New York Times*. The technology was all there. All he had to do was go downstairs, lay out the letters, ink up the typeface, and print away. There was nothing to stop him.

As it happens, it would take another four hundred years before the *New York Times* hit the street in 1851.[16] This was long after the printing press was invented. What was lacking was not the technology; the printing press of 1851 was not all that different from the printing press of 1450. What was lacking was the vision—the concept

---

16 "The New York Times," Wikipedia, https://en.wikipedia.org/wiki/The_New_York_Times.

of a printed newspaper, the concept of news as something that could be made tangible, consumable, and, most important, monetize-able.

As long as there has been human society, there has been news, at least in the generic sense. From village gossip to the latest rumor on the most recent battle, there was always information spread, most often by word of mouth. This desire to share information is very deep within us. It is part and parcel of our innate fascination with storytelling. News is just another kind of story, and very much one deigned to teach. It is no wonder, then, that social media, the marriage of this innate need to share with a technology that makes sharing simple, took off so quickly and rapidly became so powerful.

But once news becomes a commodity, it becomes subject to processing, packaging, and manipulation. Today, almost all news is "fake news." It is not fake in the sense that it is a lie, though some of it certainly is; it is fake in the sense that its significance, magnified through the media, is more often than not out of proportion to its real importance. There are perhaps a thousand newsworthy things that take place every day somewhere in the world, yet the news media in aggregate selects but a paltry handful to report to you. And, remarkably, they all report more or less the same stories.

The act of reporting those few stories inflates their significance in your mind. *These must be really important things*, you say to yourself, as every major news organization is talking about them. *Why else would it be on the news?* So you are compelled to pay careful attention.

That selection of a paltry few stories of note is then magnified a hundred thousand times through the echo chamber of the Internet and social media. The import is not in the stories per se, but rather in the significance that they are suddenly given by dint of their selection, and by the same token, those which are ignored seemingly have little to no value.

Marry the concept of news to the highly attractive and addictive technology of video, television, web, or otherwise, and you have a recipe for a very powerful medium, and one with very few controls. This technology becomes our vehicle for not only learning about the world but also shaping the decisions we will make.

FOLLOW THE MONEY

To understand how all this works, and most important, how you are being affected by it, you must first understand that the news business is, first and foremost, a business. The commodity that it is selling is not information. The commodity that it is selling is you.

The primary purpose of the news business is to make money for the corporations that own the news companies. The news business, and hence journalism, is not some holy order, dedicated to delivering the truth to you every night, nor is it in the business of educating you in order to make you a better citizen. It is, like every other media company in the world, in the business of attracting as many eyeballs as possible and holding them long enough to sell those eyeballs or that data to advertisers. The more eyeballs they attract and hold, the more money they make. What they, the news companies, decide to present to you as news is entirely a function of what will garner the most eyeballs, and nothing more. That and that alone is what drives their decisions as to what to cover and what to ignore. And bear in mind, they end up covering about 0.1 percent of what actually goes on in the world, ignoring the other 99.9 percent. If they don't cover it, then for all practical purposes, it never happened. You are not getting a true picture of the world; you are getting one that is heavily mediated.

As it turns out, what will garner the most eyeballs are scary stories, things that will frighten you. The more the news organizations and the media companies that own them can scare you, the more you will be compulsively driven to watch. This is why the news is endlessly filled with stories about terrorists, war, danger, death, destruction, and disease, with the occasional human-interest story thrown in at the end, along with sports and weather.

This would all be fine if we all realized that the news business is merely there to entertain us, but we don't. We live under the mistaken impression that there is some fundamental greater value to news, that we can learn something from it and, from what we have learned, create intelligent opinions on important topics and then act on those learned opinions. None of this is true.

All that entertainment with scary stories does, however, have an impact. Twenty-four-hour cable news and the never-ending Internet feeds like Twitter and Facebook have only magnified the problem. But what happens to people who are endlessly bathed in scary stories that they believe to be true? What happens to a society that does this to itself day after day for years on end? Do most people end up living in some kind of ongoing anxiety? And if they do, what does that do to them over time? Does it warp the way they perceive the world, their own personal safety, their relationship with others, the people for whom they vote?

I think so. As a result, I think it is critically important to understand exactly how this very powerful machine actually works.

WHEN NEWS BECOMES THE BUSINESS OF JOURNALISM

The Graduate School of Journalism at Columbia University, from which I graduated and where I later taught, was not founded until 1912, a good 2,500 years or so after Pheidippides delivered his own news. The school was founded and funded by Joseph Pulitzer, of the Pulitzer Prizes for Excellence in Journalism. When Pulitzer offered Seth Low, then president of Columbia, money to found such a school, the university turned him down. News and newspaper work were not viewed as valid subjects for a university, not an academic discipline. The journalism school was not actually founded until after Pulitzer's death, when he left the university a $2 million bequest for such a purpose. When it comes to academic standards versus money, money generally wins out. In 1917, the school awarded the first Pulitzer Prizes in Journalism.

Joseph Pulitzer was born in the Hungarian city of Makó in 1847. His father, Fülöp Pulitzer, who had been a successful merchant, died at an early age, leaving Joseph a business that soon went bankrupt. Without a way to earn a living, Pulitzer attempted to enlist in several European armies, but without success. He found better opportunities with Massachusetts recruiters who were seeking soldiers for the American Civil War and paid for the 17-year-old to come to the United States and fight on the Union side in 1864. Pulitzer eventually was part of the Lincoln Cavalry from New York and fought with General Sheridan.

After the war, Pulitzer made his way to St. Louis, where he held a number of odd jobs, including working as a waiter. After recording land rights for the railroad, he decided to attend law school. But because of his broken English, Pulitzer found the legal profession difficult, even if he did speak fluent French, German, and Hungarian. Instead, he got a job at the *Westliche Post*, the German-language newspaper in St. Louis.

Pulitzer turned out to be an excellent newspaperman. He worked tirelessly and soon got the moniker "Joey the Jew." Ultimately, he saved enough money to buy a small English-language paper, the *St. Louis Post*, and then the *St. Louis Dispatch*, and merged the two papers together to form the *St. Louis Post-Dispatch*, which is still in business to this day. The paper proved to be remarkably successful, and Pulitzer, in his spare time, also got elected to the state legislature. He was the very living definition of an upward aspiring and remarkably successful immigrant—rich, successful, aggressive.

In 1883, Pulitzer, by then a wealthy man, purchased the newspaper the *New York World* from Jay Gould for $346,000. The paper had been losing $40,000 a year, and Gould was glad to be rid of it. But Pulitzer had some rather novel ideas on how to make it profitable. Instead of doing traditional news, he filled the paper with sensationalized stories, crime, murder, disasters, and scandal. In 1895, he added the first color comics, which also proved remarkably popular. Pulitzer had an innate understanding of what the general public wanted to read, and it wasn't news, at least not in the conventional sense at that time. And it worked. Pulitzer's circulation grew from fifteen thousand to six hundred thousand, making it the largest newspaper in the country. And, of course, as his circulation grew, so did the number of advertisers and the revenue that they provided him with.

Not to be outdone, his primary competitor, William Randolph Hearst, saw the writing on the wall and first took his San Francisco–based paper, the *San Francisco Examiner*, downmarket, to great success, and later purchased the *New York Journal*, providing Pulitzer with a competitor in the world of lurid crime, sex, scandals, and the sprinkling of occasional nudity (such as it was at the end of the nineteenth century). The "press wars" between the two papers led to a massive downward spiral in the quality of the journalism being delivered, while their revenues spiraled upward accordingly.

The bottom of the barrel, more or less, was reached with the advent of the Spanish American War, a war that some have dubbed "the Journals' War," so aggressive was Hearst in using the conflict to advance the sales of his papers. When the battleship the USS *Maine* exploded and sank in Havana Harbor, the *Journal*, with no evidence whatsoever, blamed the Spanish government, and the call to arms, "Remember the *Maine* and to Hell with Spain" became a national rallying cry. A few months later, the United States was indeed at war with Spain.

Hearst also famously sent artist Frederick Remington, who is perhaps better known for his paintings and bronzes of the American West, to Cuba. Finding little to draw, Remington telegraphed Hearst, "Everything is quiet. There is no trouble. There will be no war. I wish to return." Hearst sent back a note: "Please remain. You furnish the pictures and I'll furnish the war." And he was as good as his word.

The upshot of the Hearst/Pulitzer rivalry and the product it produced in such vast numbers, not to mention the astonishing wealth it created for both men (the film Citizen Kane is based loosely on Hearst), has been labeled "yellow journalism," thought to be derivative of one of Pulitzer's early cartoon characters, the Yellow Kid. Today, this type of journalism is instantly recognizable as the mainstay of pretty much every news organization in the world. That is how they sell papers; that is how they get people to watch their TV news shows; that is how they get likes and eyeballs on their online newsfeeds. The inherent irony, of course, is that most of this was invented by Joseph Pulitzer, whose name and prize are synonymous with quality journalism.

Today, thanks to TV, cable, satellites, Facebook, Twitter, Instagram, and so much more, we are awash in this kind of journalism twenty-four hours a day. The great mistake we all make is that we take any of this seriously. Some decry what appears on TV or in the press as "fake news." It is *all* "fake news" in that it is not really about news—it is about entertainment.

Contrary to what we were taught at Columbia, news and journalism are not some kind of sacred trust dedicated to endlessly finding out and reporting the truth; they are a business, dedicated to maximizing revenue by maximizing audience, and doing whatever it takes to get there. And what it takes, what works, is offering the people what they want to see, not what they should be seeing. The news is nothing but the ultimate reality TV show. It takes reality and turns it into both entertainment and revenue, all at your expense.

If you turn on your local news, you will, with a very few exceptions, see that every evening, no matter what is happening in the real world, the stories on the news will all be about the same: a murder, a fire, an automobile accident, and, after sports and weather, a heartwarming story about a soldier who came back to town, unannounced, and suddenly appeared at his kid's fourth-grade class. Almost all local TV news looks exactly the same, no matter what town or city you live in. It's a formula, and it has not changed since the days of Joseph Pulitzer and William Randolph Hearst. What gets reported as the news, regardless of what platform, is almost entirely a function not of the value of the story per se, but rather its potential to capture and hold an audience.

Of the top stories appearing on your TV station—the murder, the fire and the robbery—not one of those stories has any impact on anyone who lives in your city or town, unless, of course, you happen to be, or be related to, the poor bastard who has been murdered, whose house has burned down, or who was robbed. As John Ford, who has run many cable networks, explained to me, television is like video flypaper. The object is to catch people who are buzzing past and hold them.

Researchers at the Max Planck Institute for Human Cognitive and Brain Sciences in Germany conducted a series of studies in which researchers exposed six-month-old babies to images of spiders and snakes, as well as flowers and fish. As it turned out, babies who had had no prior exposure to or knowledge of either spiders or snakes showed immediate elevated levels of anxiety and fear with the spider and snake images, but none with the flowers or fish images. This was done by measuring pupillary response, which indicates levels of the fight-or-flight chemical norepinephrine.

The researchers concluded that our fear of snakes and spiders is not learned but rather genetic and instantly instinctive. Wrote Dr.

Stefanie Hoele, lead investigator for the project, "We conclude that fear of snakes and spiders is of evolutionary origin." It is that same evolutionary anxiety, that same fear, that makes you jump when you see a spider or a snake—that rush of adrenalin that media companies in general and the news business in particular taps into. It grabs your attention. You cannot resist. You are preprogrammed to respond.

It is not just our innate arachnophobia that is being tapped into, but it is also the entire range of what was once a valid fear, now simply buried but still very real. We may be on the constant lookout for snakes and spiders, but we are also acutely aware of the innate dangers presented by our fellow man.

Until fairly recently in human history, a short walk down a forest path from one village to another was the spider equivalent of taking your life in your hands. The open road and certainly the forest was filled with brigands ready to pounce on you, beat you if not kill you, and take your meager copper coins. And if the brigands didn't get you, the endless wars and roving armies certainly did. For almost as long as there have been human beings, there has been violence and danger on every corner, ironically, until now.

So fear is deep within us, particularly fear for our own personal safety. It has been bred into us for about a million years. It is instinctive, like the craving for fat and sugar. It triggers an immediate response. And based on our history and experience on this planet, it is not unreasonable. Only those who properly feared death lived long enough to reproduce. If you are alive today, it is only because your grandparents, and great grandparents, a hundred thousand generations back, all of those beings were responsive enough to fear for their own personal well-being and lived long enough to reproduce. Otherwise you would not be here. So the drive for survival, the sensitivity to danger of any kind, is deep and fundamental. You don't even have think about it. It is there.

The media companies understand this and tap directly into it, that basic fear and hypersensitivity to danger. We are on the constant lookout for our own survival, even if the truth is that the world has, ironically, never been a safer place. There are no more saber-toothed tigers looking to make a meal out of you, no more flashes of light in the tall grass of the African savanna indicting the presence of a lion, no more wandering bands of brigands with knives and swords ready to cut you down for your purse of a few copper coins, and no more roving armies of your arch enemy out for blood. The danger is gone, but the sensitivity to it is still there. And that is your Achilles' heel in the eyes of the media companies. That's their opening. And they take it.

EBOLA!

A mysterious virus that you cannot see brings on an agonizing and painful and terrible death where your internal organs actually liquify inside you and you bleed out through every orifice including your eyes. You could not write a more terrifying science fiction script, but it was real! It was a ratings killer.[17] It still is to this day.

And so the evening news and every other news organization, from time to time, becomes a never-ending stream of Ebola stories that could scare the pants off of you and keep you tuned it. That, of course, is the whole idea. And people watch and respond. They resonate with the appropriate fear and incorporate that fear and anxiety into their everyday thinking and actions. Ebola is the spider in your bed—terrifying! And it could happen to you!

Now, as it turns out, during the last great Ebola epidemic in Africa, 3,400 people died from the disease (no one in the United States, by the way). That's not great. However, according to the CDC, 2,195 children die every day from diarrhea; that's 801,175 deaths a year, more than the combined total of AIDS, malaria, and measles.[18] But does diarrhea make the evening news? I think not.

In Strong, Maine, for example, the last time one of these Ebola scares rolled through town in 2014, an elementary school teacher was put on leave for twenty-one days because she had traveled to an educational conference in Dallas and stayed at a hotel that was 9.5 miles

---

17 I can attest to this personally. When I was running New York Times Television, I produced *Killer Virus* for TLC (all about the Ebola "danger") and won a national Emmy for News and Documentaries.

18 "Global Diarrhea Burden," Centers for Disease Control and Prevention, https://www.cdc.gov/healthywater/global/diarrhea-burden.html.

from a hospital where two nurses had contracted the disease.[19] "At this time, we have no information to suggest that this staff member has been in contact with anyone who has been exposed to Ebola," the school's administrative office said. "However, the district and the staff member understand the parents' concerns. Therefore, after several discussions with the staff member, out of an abundance of caution, this staff member has been placed on a paid leave of absence for up to 21 days."

In Georgia, a school district closed enrollment to students from Liberia, Sierra Leone, and Guinea unless they can present a doctor's clean bill of health. In Hazelhurst, Mississippi, parents withdrew their children from a middle school after learning that the principal had been to Zambia for his brother's funeral. Zambia, just a country away from South Africa, is well over two thousand miles away from the Ebola outbreak in West Africa.[20]

Syracuse University disinvited a Pulitzer Prize–winning journalist from speaking because he recently went to Liberia for work.

The fear was palpable and the media coverage extensive and terrifying. And the number of Americans who actually died of Ebola contracted in the US? Zero.

Now, the number of Americans who died in automobile accidents in the same year? 32,675.[21] But death in an automobile accident is not scary, not really—not as scary as having your organs liquefy while you are still alive and not being able to do a damned thing about it. That's scary. So it becomes a leader in the news business.

Amos Tversky and Daniel Kahneman, the latter a Nobel Prize–winning author of *Thinking Fast and Slow*, noted a phenomenon they called the "availability heuristic." According to them, people categorize the probability of an event, not by the actual statistical likelihood

19 Caroline Ferguson, "Ebola Scares in Maine," Colby Echo website, October 23, 2014, http://colbyechonews.com/ebola-scares-in-maine/.
20 Alan Yuhas, "Panic: The Dangerous Epidemic Sweeping an Ebola-Fearing US," *The Guardian*, October 20, 2014, https://www.theguardian.com/world/2014/oct/20/panic-epidemic-ebola-us.
21 "2014 Motor Vehicle Crashes: Overview," National Highway Traffic Safety Administration, March 2016, https://crashstats.nhtsa.dot.gov/Api/Public/View Publication/812246.

of that event, but rather by the ease and rapidity of which that event comes to mind. Things that are recent, upsetting, violent, graphic, or shocking tend to come up first in the memory. Take plane crashes. The odds of being killed in a plane crash are 0.06 per one million flights, or one fatal accident for every sixteen million miles flown. The odds of dying in a car crash are 1:103.[22] Yet 60 percent of Americans report that they are afraid to fly. Hardly anyone is afraid to drive in a car. Americans rank tornados as a more common cause of death than asthma. Tornados kill about fifty Americans annually, while asthma kills about four thousand. But tornados make better TV.[23]

ISIS, like Ebola, is a ratings winner. The fact that ISIS made its own videos of its beheadings (and other executions) only made it easier, and in some strange way more intriguing, for the media companies to play them up as a major world threat. Cutting off someone's head is even more frightening than being in a plane crash. It ranks very high on the Tversky/Kahneman scale of things that stimulate your mind.

While public beheadings are certainly gruesome and terrible affairs, it is interesting to note that in 2018, ISIS publicly beheaded 65 people, while Saudi Arabia publicly beheaded 110.[24] Those Saudi beheadings don't make the news, however. And no presidential candidate, at least not yet, has called for the carpet-bombing of Saudi Arabia, nor promised to go to Congress to get a congressional declaration of war against Saudi Arabia.

ISIS is the Ebola epidemic after the Ebola hysteria dies down. News executives always need another threat waiting in the wings. People have short attention spans, and they soon grow tired over the

22 Jessica Bursztynsky, "Americans More Likely to Die from Opioid Overdose Today Than Car Accident," CNBC website, January 16, 2019, https://www.cnbc.com/2019/01/15/americans-more-likely-to-die-from-opioid-overdose-than-car-accident.html.

23 Daniel Pinker, "The Media Exaggerates Negative News. This Distortion Has Consequences," *The Guardian*, February 17, 2018, https://www.theguardian.com/commentisfree/2018/feb/17/steven-pinker-media-negative-news.

24 Megan Specia and Vivian Yee, "Saudi Teenager Faces Death for Acts When He Was 10," *New York Times*, June 9, 2019, A4, https://www.nytimes.com/2019/06/09/world/middleeast/saudi-teenager-death-sentence.html.

"threat" of the moment. News stations will lose their viewers if they stick too long with one fear. They have to rotate anxiety-inducing stories.

Since 9/11, 45 Americans have died as a result of Islamic terror attacks in the US, and 104 worldwide.[25] That's bad, but hardly a disaster, really. Yuval Noah Harari wrote in his recent book *Homo Deus,* "In 2012 about 56 million people died throughout the world: 620,000 died due to human violence (120,000 were killed in wars). In contrast, 800,000 committed suicide and 1.5 million died of diabetes." Yet, in the US we spend $16 billion a year fighting terrorism[26] and just over $1 billion a year fighting diabetes, according to the *Journal of the American Medical Association.*[27] That's about $360 million per terror victim. By way of comparison, we spend about $38 per person with diabetes in search of a cure. Terror attacks make far better TV than does diabetes research.

News stories of someone opening up with a semi-automatic rifle and killing a dozen people in a church or a school or a parking lot somewhere in America (an increasingly fairly standard story) may get ratings, but the truth is that, as terrible as this may seem, a dozen people getting shot in an afternoon represents something that happened to 0.000003 percent of the country. In other words, that is statistically unimportant. Does this seem cold? Perhaps. Or is it perhaps colder to scare the bejezus out of everyone every night so that you may sell more Cialis?

On average, nearly 790,000 Americans die of heart disease every year.[28] By contrast, 74 Americans on average die from terrorist incidents every year. How frightened are you of being caught up in a terrorist event? How frightened area you of your KFC hot wings? Which

25 Peter Bergen et al., "Terrorism in America after 9/11," New America website, https://www.newamerica.org/in-depth/terrorism-in-america/.

26 Drew DeSilver, "U.S. Spends over $16 Billion Annually on Counter-Terrorism," Pew Research Center, September 11, 2013, https://www.pewresearch.org/fact-tank/2013/09/11/u-s-spends-over-16-billion-annually-on-counter-terrorism/.

27 Kelly Close, "JAMA Paper Breaks Down Medical Research Funding in the US," DiaTribe Learn website, https://diatribe.org/jama-paper-breaks-down-medical-research-funding-us.

28 "Stay Aware of Heart Disease," Boston Scientific website, https://www.your-heart-health.com/content/close-the-gap/en-US/heart-disease-facts.html.

is actually more likely to kill you? KFC buys lots of TV commercials; terrorists, not so many.

The truth is that if you are like most Americans, you are somewhere between concerned and terrified of terrorism, even if the odds are that you will never in your life come close to one of these things. You are, meanwhile, piling on the hot wings and the bacon. Really, what are the odds that you are going to be caught in a terrorist action? According to the Life Insurance Institute, the odds on your being killed in a terrorist attack are one in twenty million. In other words, it is statistically almost impossible. And if that were not sobering enough, the odds on being killed by a foreign terrorist, according to Congressman Ted Lieu, are 1:3.6 billion—in other words, highly unlikely. Yet how terrified are we of being attacked by terrorists? The answer: very terrified. However, as my friend Gary Younge points out in the *Guardian*, more Americans were shot by toddlers last year than by ISIS. How frightened are you of toddlers?

We are terrorized, all right, but we are more terrorized by the news media than we are by any real-life terrorist. Those terrorists, when they do exist, are far away and a rarity. The "news terrorists" are in your living room and on your phone all the time, terrifying you. But that is their job. Their job is to get you to watch and keep watching. Watch TV, read the papers, listen to the blogosphere, and you would think that the nation was on the verge of collapse.

There are important stories in a community that do impact people—sewer bond issues or educational issues—but those are boring. They don't "rate," so they don't make air. But a cumulative diet of shootings, robberies, fires, and murders every night has an effect on you. Over time, it makes you believe that you live in a very dangerous neighborhood.

Ironically, if you were to chart the history of human violence, as has Dr. Steven Pinker, you would see that, contrary to what TV might be telling you, we are now living in the safest and most peaceful period in all of human history. If a news executive put that on the air, the ratings would crash.

In 1965, Johan Galtung, a Norwegian professor, carried out what might have been the first critical evaluation of the impact of news on the public.[29] "The consequence of all this is an image of the world that gives little autonomy to the periphery but sees it as mainly existing for the sake of the centre," Galtung wrote more than a half century ago. "Conflict will be emphasised, conciliation not." And this is precisely what has happened, Galtung said. "News media give a total biased picture of reality. The perception of reality in the public becomes overly negative." Some fifty years after its original publication, Galtung's opinions had not changed: "I was saying, 'What you do is incomplete. You are missing a major part of the image of the world.'"

People are naturally primed to keep an eye out for pending disaster. There is a natural proclivity for it. We are drawn to it as moths to a flame. When there was an actual danger afoot, it was a fantastic survival mechanism. But when there is no danger, or at least no danger inherent in the kinds of stories that the news broadcasts ad nauseam, the danger described may drive the ratings for the TV and cable channels and bring more clicks online, but cumulatively, it has a deleterious effect on the individual who is constantly exposed to it. It makes us anxious, depressed, unhappy, hopeless, and constantly frightened.

FOOD FOR THOUGHT

Until about a century ago, almost everyone in the United States, and for that matter, everyone on earth, was involved in growing food. Agriculture was our number one employer. It was pretty much our only employer, and this was not out of choice. People living three thousand years ago did not agonize about what they were going to do for a living when they graduated from university. It was not because they all liked farming; it was a matter of simple survival. If they could not tease enough food out of the soil, they starved to death. A bad turn in the weather, a cold snap, a drought, a flood, and death stared them in the face. This had been the human condition since humans planted the first seeds some twelve thousand years ago.

The rise of industrial agriculture and farming suddenly separated us from our direct relationship to food, and thus it allowed the rise

---

29 Nils Petter Gleditsch, Jonas Nordkvelle, and Håvard Strand, "Peace Research: Just the Study of War?" *Journal of Peace Research* 51, no. 2 (March 1, 2014): 145–158, https://journals.sagepub.com/doi/full/10.1177/0022343313514074.

of corporate food companies and Twinkies and every other unhealthy and chemical food we now consume. This could not have happened in an era in which we had so direct a connection to how food was grown and created. Our current relationship to our food, or rather our lack thereof, is slowly killing us.

When it comes to media, it is very much the same. We are, all of us (or at least 99.999 percent of us), vastly separated from the way in which media is made. We always have been, for the past seventy years or so, separated from media since the rise of the industrial media complex, coincidentally about the same time we lost touch with food. When Pheidippides ran from Marathon to Athens to announce the victory, everyone understood what had happened. Everyone could relate, on a very basic human level, not just to the news but also to the way in which it had been delivered. It was comprehensible. This is no longer the case.

For us, media and information simply appears, the way Pringles does, on the grocer's shelf. Almost no one understands or is in any way directly connected to the way that it is made. As a result, the media conglomerates can make it any way they want. What do we know? What do we care? Is it nutritious? Who knows or cares? Is it good for us? Ask a network news executive and he or she will think you are out of your mind. The object of what is put on television or Netflix or your Facebook feed is something that will "rate," something that you will consume. Whether it is any good for you is a question that does not even cross the mind of those in corporate media. If you worked in the industry and you asked that question, you would soon be shown the door. You clearly would not understand what the media business is all about. It is not about imparting any kind of knowledge; it is about selling the spaces around the content.

On October 15, 1958, Edward R. Murrow, perhaps the best respected of any television news journalist in the short history of the medium, was asked to speak at the annual meeting of the RTNDA, the Radio Television News Director's Association, the national organization that represented his industry, and an industry in which he had made his career and fortune. Even if you never heard of Murrow, you would recognize him immediately. He became the role model for every on-air reporter and correspondent you have ever seen. The wall at the entrance to the CBS Broadcast Center on Manhattan's West Side is emblazoned with his picture.

At the RTNDA meeting that year, Murrow had been expected to deliver the usual pap about how great the industry was. Instead, he took the media to task for failing to deliver on what television might have been, but was not. "This instrument can teach, it can illuminate; yes, and it can even inspire. But it can do so only to the extent that humans are determined to use it to those ends. Otherwise it's nothing but wires and lights in a box."

By 1958, the then very young TV networks were at a crossroads. Would TV be used to educate and illuminate, or would it be used for mindless entertainment in a never-ending race for ratings at any price? Murrow not only saw what was coming, he lived it. And, even though he had made his career and his fortune and fame from TV, he read out this warning at the biggest yearly meeting of television executives in the country, the RTNDA.

To fully comprehend both the bravery and the astonishing reaction his speech must have elicited, you have to understand that for the fast-growing television industry, Murrow was something of an icon. For him to speak the truth about the business was, in effect, as though Warren Buffet suddenly announced that he was, in fact, a Communist and that capitalism was the source of all of mankind's problems. It was electrifying.

For this, Murrow would be roundly castigated as "fouling his own nest." In fact, it was the end of his career. That is courage.

Ed Murrow, like Lincoln, was born in a log cabin, in 1908 in a Greensboro, North Carolina, near the Polecat Creek. This was a time before electrification, before anyone had a telephone, before radio. To find a place today as comparably isolated from the rest of the world as was his birthplace, you would have to start on Mars. When Murrow[30] was six years old, his family moved to a homestead in northern Washington State, about sixty miles from the Canadian border. You can't get more rural than that, even by Polecat Creek standards. He

---

30 A. M. Sperber, *Murrow: His Life and Times* (New York: Fordham University Press, 1999). This book is probably the definitive biography of Ed Murrow.

went to nearby Edison High School, and by his senior year was elected president of the student body and was on the debate team.

Murrow enrolled in Washington State College in Pullman, which was already a leap for someone born in a log cabin, and there he majored in speech. He also got very involved in student government, and as such, ended up as a speaker at the National Student Federation of America meeting in 1929. He gave a speech there, and it was so good that he was immediately elected president of the federation.

From his humble beginnings in North Carolina, the first in his family to attend college, Murrow had already achieved great success, but he had barely gotten started. In 1930, Murrow moved to New York and was hired as the assistant director of an organization called the Institute of International Education. It was the job of the institute to try to find work and placement for the hundreds and later thousands of German intellectual and academic refugees who were fleeing the Nazi takeover in Germany. This gave the young Murrow enormous exposure both to events happening in Europe on a very personal basis and to amazing European thinkers and leaders.

In 1935, Murrow left the IIE and joined CBS, which was a fledgling radio network, as the director of talks and education. Today, we think of CBS as a media giant, which it is, but in 1935, it was only eight years old and still struggling to get on its feet. In 1927, twenty-six-year-old Bill Paley, son of a cigar baron, bought a group of sixteen little radio stations and strung them together to create a network. It was tough going, and the Great Depression of 1929 did little to help the bottom line. However, Paley made it work.

In 1935, CBS's entire news staff consisted of but one man: Robert Trout. Murrow's job was to line up speakers for Trout to interview on the radio. In 1937, Murrow was sent to Europe, and his job was to convince European leaders to speak on the CBS radio network. NBC was the dominant radio network at that time, having not just one network, but two: NBC Red and NBC Blue, which would later become ABC. CBS was a tiny competitor of NBC, but Murrow worked hard to make CBS a success. As it happened, one of the speakers Murrow lined up was a member of Parliament from Epping, a young Winston

Churchill. It would turn out to be a good choice and the beginning of a good relationship for both.

At first, Murrow worked alone. However, after a few months he was allowed to hire one assistant, the journalist William L. Shirer. Shirer had been in Europe since 1925, working for the *Chicago Tribune*. Those were the days when newspapers were dominant and had correspondents all over the world. Newspapers also had big budgets, and Shirer had travelled across Europe, the Middle East, and India for the *Tribune*. He had even befriended Mahatma Gandhi. Radio work was considered a big step down.

In 1935, Shirer shifted over to working for the Universal Service, one of William Randolph Hearst's two wire services. When Universal folded in 1937, Shirer was moved over to Hearst's other service, the INS, or International News Service. When that went broke a few weeks later, Shirer, who would pun that he had gone from "bad to Hearst," found himself out of work for the first time in a dozen years. As it happened, Murrow was looking for someone to help him out with his increasingly large workload. Shirer fit the bill.

As Europe moved inexorably toward war, there was more and more interest in booking Murrow's speakers for CBS radio. Today, having been bathed in electronic media, we tend to forget that in its early days, the electronic media were looked at as a kind of joke, certainly when compared with the power and prestige of newspaper or the wire services.

The decision to hire a newspaperman at that time struck Murrow as odd and perhaps a mistake. In those days, the correspondents, like Murrow and now Shirer, were actually forbidden from broadcasting on the radio. They were seen a mere fact gatherers on the ground. The real reporting was done by the professionals from the studios in New York. (Interestingly, for many years then and thereafter, *Time* magazine, considered the apex of print journalism to many, was produced in exactly this way, a process that would continue well into the 1990s.) For CBS, all the information was gathered in the field, but it was from New York that the actual broadcast was generated.

Both Murrow and Shirer felt that CBS's prohibition on correspondents broadcasting from the field made no sense, and in 1938, they got their chance.

HITLER AND MURROW AND THE NEWS

On March 12, 1938, Hitler's 8th Army crossed into Austria. For a long time, Hitler had been pressuring both Austria and the rest of the world for the uniting of Germany and Austria as one German-speaking country. This was called Anschluss. Hitler rolled the dice and went ahead. He and his generals were terrified that the British and French would send in their own troops to stop Hitler. They did not, and it was Hitler's first great victory, and a bloodless one. When the Wehrmacht entered Austria, it was actually greeted by cheering crowds waving German swastikas. A few hours later, Hitler crossed into Austria near Braunau, his birthplace.

The enthusiasm with which the Austrians greeted the Germans and Hitler was a surprise, both to Hitler and to the rest of the world. Germany, as a country, had only come into existence in the 1860s, crafted by Otto von Bismarck. Before that, it had been an amalgam of tiny states and principalities. Austria had competed with Prussia to unify the German-speaking peoples, but it had been Prussia that had proven dominant. Now, Austria had been brought into the German fold. The Anschluss had come off far better than even Hitler imagined possible. Had the British and French invaded Germany, it would have been the end for both Hitler and the Nazi Party.

As the tanks rolled into what had been Austria, the border barriers came down, and the western part of Austria was incorporated into Germany proper. It was an electrifying moment, both for Germany and for the world. The terrible memories of the First World War were still raw in everyone's mind.

Having crossed into the former Austria, Hitler decided to continue his 'tour' of his newly conquered land, and on March 15, he entered Vienna, and at the Heldenplatz declared to a crowd of two hundred thousand cheering supporters: "Als Führer und Kanzler der deutschen Nation und des Reiches melde ich vor der deutschen Geschichte nunmehr den Eintritt meiner Heimat in das Deutsche Reich." ("As leader and chancellor of the German nation and Reich, I announce to German history now the entry of my homeland into the German Reich.")

Hitler had just pulled off one of the biggest coups in European history. Emboldened, Hitler would go on to absorb the rest of Austria, and later Czechoslovakia, and almost all of Europe. But that was in the future.

On that March day, the Fuhrer's arrival in Vienna was the biggest news story in the world. Ed Murrow wasn't there. He was in Poland trying to line up a children's chorus for a live performance on the CBS Network. Although Shirer was in fact in Vienna and was able to tip Murrow off about what was happening, the Austrian officials at Austrian radio were preventing Shirer from using their broadcast facilities to relay information back to New York. On his own volition, Murrow chartered a twenty-three-seat airplane (in 1935, plane travel was still considered pretty rare), and flew to Austria to cover the arrival of Hitler's armies (and Hitler) live for CBS. Contrary to CBS's rules and tradition, Murrow reported live from Vienna to an eager American audience: "This is Edward Murrow speaking from Vienna. It's now nearly 2:30 in the morning, and Herr Hitler has not yet arrived."

It was riveting and revolutionary. And no one had ever done anything like that before. CBS management was so taken that they authorized both Murrow and Shirer to expand the reportage, to which they gave the title, *World News Roundup*. *World News Roundup* is still on CBS radio to this day, making it the single longest running program in broadcasting history. It was also the launch of Murrow's career as a journalist.

As the war heated up in Europe, Murrow and Shirer's reports became required listening in American households. When Germany began its intense bombing of London after Britain's entry into the war in 1939, Murrow became a household name.

With the Luftwaffe pounding London and indeed all of England from the air in September 1940, Murrow, now the chief European correspondent for CBS, had been seeking permission from the British Air Ministry to broadcast live from London's rooftops. In fact, there was a stricture at CBS and at NBC against the airing of any kind of recorded broadcast. Everything had to be live.

In any event, the Air Ministry was concerned that by broadcasting a live report on the German bombing, the Germans might in some way gain important information as to the location of their bombs being dropped. In those days, bombing itself was a kind of hit or miss event. To demonstrate his idea, Murrow and a sound technician went to the roof of the BBC Broadcasting House and recorded a demonstration for the Air Ministry of what they would say. Again and again, six times actually, he was turned down.

But Murrow was not to be deterred. Finally, he went directly to Churchill, his old friend from the past, who gave permission. On the 21st of September, 1940, Murrow began what would become one of the most famous broadcasts in radio history:

> This [dramatic pause] is London. I am standing on a rooftop looking out over London. At the moment everything is quiet. For reasons of national as well as personal security, I'm unable to tell you the exact location from which I'm speaking. Off to my left, far away in the distance, I can see just that faint, red, angry snap of antiaircraft bursts against the steel-blue sky, but the guns are so far away that's impossible to hear them from this location. I think probably in a minute we shall have the sound of guns in the immediate vicinity. The lights are swinging over in this general direction now. You'll hear two explosions. There they are! That was the explosion overhead, not the guns themselves. I should think in a few minutes there may be a bit of shrapnel around here. Coming in, moving a little closer all the while.

Punctuated by the sound of German bombs going off, the firing of antiaircraft guns, the droning of the Luftwaffe overhead, and the wail of British air raid sirens, no one had ever experienced anything like that before. Churchill had made the right decision. Murrow's nightly broadcasts from the roof of the BBC would greatly affect America, which was still officially neutral, and sway public opinion on the side of the British.

The medium was new, and the experience of live broadcasts as German bombs fell on London and air raid sirens screamed in the background was absolutely riveting. When Murrow returned to the United States in 1941, CBS had a welcome-home dinner for him at the Waldorf-Astoria Hotel in New York. Eleven hundred people attended. President Franklin D. Roosevelt sent a telegram of congratulations. Murrow was the first media celebrity.

Then, four days later, the Japanese bombed Pearl Harbor, and Murrow became the voice of America's media. He worked so closely with the British that in 1943, Winston Churchill offered to make him a director of the BBC, a job he turned down. During the Second World War, Murrow continued to report for CBS, flying along on more than twenty-five combat missions over Europe and greatly expanding CBS's London bureau, hiring some of the best and most talented print and wire reporters and taking them into radio. They were called "Murrow's Boys," and Bob Pierpoint, with whom I worked at CBS many many years later, was one of them.

On April 2, 1945, Murrow and another CBS reporter, Bill Shadel, were the first journalists to enter the Nazi concentration camp at Buchenwald. In his radio report, three days later, he said, "I pray you to believe what I have said about Buchenwald. I have reported what I saw and heard, but only part of it. For most of it I have no words. . . . If I've offended you by this rather mild account of Buchenwald, I'm not in the least sorry."

When television came along, Murrow was dragged, kicking and screaming, into the new medium and became its very face. Today, if you walk into the CBS production studios on Manhattan's west side, you will see the visage of Murrow staring down at you.

Murrow's producer in television was a moose of a man named Fred Friendly, who, as it turns out, was my mentor and teacher in the television business. Together, Murrow and Friendly used the new medium of television to do amazing things, things that demonstrated the enormous potential of the new medium. At their apogee, they brought down Joseph McCarthy and his national red scare. All they did was follow McCarthy around the country, filming his speeches. Then, they cut his speeches together and aired them for television— nothing more. The man was hoisted on his own petard, and following the TV show, the nation began to turn against McCarthy and McCarthyism.

It was a stunning achievement, an astonishing demonstration of the power of this new medium. It could make people, and it could break people. It could turn the tide of a nation in a single broadcast.

Murrow and Friendly's show was called *See It Now*, and it tackled, in a very intelligent and aggressive way, some of the most serious issues facing the country. It was television journalism at its best, and some of the shows, such as "Harvest of Shame," about migrant workers and sharecroppers, are even today considered the gold standard of documentary television.

But this was not to last. Murrow and Friendly may have been doing stellar work, but their shows did not rate. At least they did not rate as well as the newly invented format of quiz shows. Now those rated, and when Murrow watched *The $64,000 Question* for the first time, he turned to Friendly and asked how long he thought they might be able to stay on the air. The answer was, not long.

By 1958, when Murrow addressed the RTNDA, he and Friendly had been relegated to the basement of CBS. The network had better ratings with the less upsetting stuff. In the end, Murrow ended up leaving CBS. He had no future there. He ended his days as the director of the Voice of America. Fred Friendly would resign as president of CBS News in 1966 when the network refused to cover the first US Senate hearings on America's involvement in Vietnam and instead reran old episodes of *I Love Lucy*.

Murrow, as it turns out, was right, and incredibly prescient. TV *did* end up as nothing but wires and lights in a box. It was at this seminal moment in the very early history of television that things could have been different. It was here that there was a parting of the ways, that the future not just of television, but of all media, was decided and defined. At that moment, the incredible power of television, which relied upon use of the public airwaves, which we all in theory own, was turned over to three individuals: Paley, Sarnoff, and Goldenson. Aside from the fact that they were handed the keys to a public goldmine, they were also given complete editorial control over what their networks would cover and air, and what political bent they would take. They had been given an enormous public responsibility, and they took it and decided to make as much money out of it as they could instead.

When television first arrived, it was transmitted analog through the air. This requited an enormous amount of bandwidth, so much so that there was only really room for three networks. The bandwidth, in theory at least, belonged to the public, and so the FCC licensed swaths of the bandwidth for the three networks—ABC, NBC, and CBS—for no cost. The quid pro quo was that those networks would, in exchange, provide public-affairs and public-service broadcasting for free. That was where the news part came in.

This all worked fine for a while until the networks discovered that the news, done in this way, was an enormous drain on their balance sheets, and that it was indeed possible to make news a profit center.[31] It simply meant tweaking the content and selling ads. The content would be driven not by what was deemed important or educational, but rather by what would rate the best. This drive for maximized ratings would drive out the early crop of great but not necessarily TV-friendly faces (mostly Ed Murrow's "boys") and replace them with a group of not-too-bright but TV-friendly "stars."

Murrow had been right. The potential of the medium had been lost, and it had instead turned what could have been an amazing machine into nothing more than flashing lights in a box. Instead of teaching us civics, citizenship, history, and an intelligent awareness of the world that would give us all the power and ability to make informed decisions, it taught us, over and over, to be afraid—very afraid.

---

31 Ken Auletta does a masterful job of explaining this in his book *Three Blind Mice*.

PART III

⊘ ⊘ ⊘

CARPE MEDIUM

THE EXTRAORDINARY POWER OF THE
MEDIA HAS COME TO DOMINATE
ALMOST EVERY ASPECT OF OUR SOCIETY,
BUT THERE IS A SOLUTION

10

The Last Gift of Steve Jobs

Having read this far, you might well be thinking to yourself, *Well, the best thing I can now do is to throw all of my screens—TV, HBO, iPhone, tablet, computer—in the trash, move into a yurt in Vermont, and call it a day.* This might be a good solution, but it is not a realistic one. Alas, unlike tobacco or heroin, you cannot just go cold turkey; you cannot break your addiction. The media is more than an addictive drug; it is also the warp and weft of our entire society, our economy, and our world.

Rather than simply suggesting that you cut the cord and move off the grid, in this last third of the book I am going to present you with, what I think, is a far more interesting and, ultimately, far better solution: take control.

THE DAY THE WORLD CHANGED

The atmosphere in the Moscone Convention Center auditorium in San Francisco was electric on January 9, 2007. The assembled were quiet, as one might be in a church, waiting with eager anticipation the appearance of the star attraction. The star attraction at this MacWorld Conference, as at every prior MacWorld Conference since the first one in 1985, was the appearance of Steve Jobs, the founder, chairman, and CEO of Apple. This was the Church of Apple, a religion for the digital age, and Jobs was more than the pope. He was Jesus Christ incarnate.

In 2007, Jobs was the unquestioned rock star of the computer world, a boy genius who would soon singlehandedly go on to create what would one day become the most valuable company in history. But more significant, his inventions, like those of Thomas Edison, continued to both capture the imagination of the nation and change the world forever.

These yearly conventions were the place where Jobs, obsessive about keeping each new device secret, would reveal the next iteration of Apple's amazing product line. What made these annual events so popular was that, unlike other companies and their rather boring computers, each of Jobs's releases was less a product presentation and more a kind of magic act, as though he had some remarkable time machine that allowed him to reach into the future and bring back things that no one could have imagined they even wanted, but that soon became an essential part of daily life. He had done this with his iPods and their connection to iTunes that revolutionized the music business; he had done it with his iMacs, with his PowerBooks. But now, on this stage, he was about to release the most revolutionary invention he and his company had ever created—and everyone in the room knew it.

Striding out on stage, lit by a single spotlight, he more than simply dominated the room. He owned it completely. Dressed in his trademark jeans, black t-shirt and sneakers, he stood silently for a moment, waiting for the deafening applause of the assembled to die down. Then, he began to speak:

Every once in a while, a revolutionary product comes along that changes everything," he began. "One's very fortunate if you get to work on just one of these in your career. Apple has been very fortunate that it's been able to introduce a few of these into the world. In 1984, we introduced the Macintosh. It didn't just change Apple; it changed the whole computer industry. In 2001 we introduced the first iPod, and it didn't just change the way we all listened to music; it changed the entire music industry. Well, today we're introducing three revolutionary products.

Jobs had never been known for his modesty. He had almost completely missed a meeting with Obama in 2010 because he insisted that Obama invite him personally. Jobs had suggested that Obama meet with several other Silicon Valley luminaries, but when the invited grew to more than Jobs had proffered, he said he now had no intention of attending at all. As Walter Isaacson noted in his seminal biography of Jobs, Bill Gates developed a grudging respect for his chief competitor, saying, "He really never knew much about technology, but he had an amazing instinct for what works."

On this January day in San Francisco, that amazing instinct would surpass even his own larger-than-life vision of himself. Dominating the stage, Jobs went on to explain that he was releasing the long-awaited and much-anticipated iPhone. But it was more than a phone. It was, as he said, three revolutionary products wrapped in one—a phone, an iPod music player, and an Internet connection. That three-in-one construction was indeed revolutionary, but Jobs himself omitted the single feature that would have an impact that not even he could then conceive of. The phone had a camera, and that would, in the end, prove to be far more significant than the phone or the music player.

Jobs's invention would very quickly become incredibly popular. In the ensuing decade, Apple sold more than one billion of his iPhones. It became the company's number one product, in fact, the mainstay of its extraordinary growth. But that was only a beginning, for competitors soon saw the incredible attraction of the phone and began to build their own copies of his creation. Soon Samsung, Huawei, Google, Sony, and many others would begin to produce and sell their own versions of the iPhone. As of now, there are more than three billion smartphones in three billion hands around the world.

But the smartphone revolution that he initiated would change the world in ways that even Steve Jobs, its progenitor, could not have imagined. For he and his competitors did not just put some three billion phones into three billion hands. What they actually did was to put three billion professional video and television production studios into three billion hands and three billion television networks into the three billion hands. It was a device that would have the potential to completely overturn the world. The original phone only had a two-megapixel camera and limited memory, but the door had been opened to an entirely new kind of future for media.

Had you gone out in 2006 to buy the video-production capability that your phone and every other smartphone has today, you would have had to spend upwards of $10 million. A broadcast-quality video camera at that time could cost as much as $100,000. Your phone not only shoots 4K video, it also edits it in the phone. Prior to 2006, a professional CMX video-editing suite could cost as much as $1 million and require a highly skilled operator. The phone also adds music, graphics, and still photos, at no additional cost.

But the biggest revolutionary change is that the phone is also a node for broadcasting. It is the equivalent of the broadcasting tower that sits on top of the Empire State Building, pumping out television signals to a paltry few million who live in its range. Your phone, married to the web, is vastly more powerful. It can reach billions at no cost.

Prior to the iPhone, even if you had made video content, there was no way immediately share it with the rest of the world unless you owned a broadcasting network or had access to a cable head-end. In 1961, A. J. Liebling, the American newspaper man, famously said, "Freedom of the press is confined to those who own one." Now, thanks to Steve Jobs, everyone owned one. In an instant, that January morning in 2007, the entire globe-spanning, all-powerful, all-consuming multi-billion-dollar media world was turned upside down.

The major media companies had been able to dominate the world of content because there was a technical and immutable barrier to entry. They owned the broadcast frequencies and licenses; they owned the cable systems and channels. They owned the access to your home. They were the gatekeepers. They would decide what you and the rest of the world would get to see, and more important, what you and everyone else would not be allowed to see. In an instant, Jobs had given access to billions of people to not just be watchers but also to be creators. It was a seminal moment in history.

But by the January 2007 presentation, Jobs was already dying. As early as 2003, he had been diagnosed with an unusual form of pancreatic cancer, known as a neuroendocrine tumor, or islet cell carcinoma. Had Jobs immediately sought conventional medical treatment, his odds on survival would have been fairly good. Instead, however, he delayed surgery, ironically preferring to scour the Internet for alternative-medicine cures. This would prove to be a fatal mistake. By 2007, the cancer had spread, and despite spending hundreds of thousands of dollars for DNA analyses, cutting edge at the time, that would tailor treatment to his specific needs, he had, in the end, waited too long.

So when he announced the release of the iPhone in 2007, his fate had already been determined. The iPhone would be his last gift, but what an amazing gift it was.

We often look up on the invention of the printing press in 1450 as a turning point in human history. And it is true that Gutenberg's invention gave everyone the power to publish, an incredible power that had, until then, been solely in the hands of the Church and the monarchy, the only ones who could afford scribes. The printing press meant anyone was now free to publish. Yet even Gutenberg's enormous invention pales in comparison to the impact that Jobs's invention would one day have. That is because, even post-Gutenberg, if you had an idea you wanted to share, whether religious, political, social, or scientific, you still had to get access to someone with a printing press, and you still had to convince that person to absorb the cost of printing your ideas or discoveries and distributing them.

What Jobs had done in a stroke was, essentially, to place a printing press in the pocket of every person on earth and deliver to them a machine that would not just print but also distribute this incredibly powerful and addictive video-based content at no cost and require almost no technical skill and the smallest of capital investments. Now that is extraordinary.

When Jobs hired John Scully, then the CEO of Pepsi, to run a just-starting Apple, he famously said to Scully, "Do you want to sell sugar water for the rest of your life, or do you want to come with me and change the world?"

The question now is, what shall we do with this remarkable tool and this newfound power?

HARLEM LIVE—IN THE BEGINNING

Today, Harlem is a hip, up-and-coming, gentrified neighborhood that everyone wants to live in. This was not always the case.

Not so many years ago, Harlem was considered a very dangerous place. "You could get killed if you go up there," my friends used to say to me. "Uptown" carried the connotation of drugs, crime, violence, and worse. And, of course, this perception was perpetuated by the media. Pretty much every TV show and movie used Harlem or uptown as a place where people did get shot, robbed, or beaten up. It was shown over and over as a place where crime was rampant: there were drug dealers on every corner, and prostitutes on every other.

Because this was Harlem's media footprint, no one wanted to go up there, real estate prices were in the gutter, restaurants sat unoccupied—except for the occasional exotic trip for an "experience" at a place like the Cotton Club or Sylvia's, with a taxi waiting outside—and of course, no one went there to go shopping. Because of its media imprint, Harlem's economy was gutted, and as a result, so were the lives of the people who lived there.

Harlem was a ghetto in every sense of the word—a racial ghetto, an economic ghetto, but also a media ghetto. Fear in the white world downtown, perpetually fed by the media and by movies—TV series driven by lurid crime stories and cops and, more than anything else, the local news media—destroyed the vast potential that Harlem had. But fear was profitable, so long as you were not a resident of Harlem or had a business there. It was the cheese in the trap that kept the viewers coming.

In response to this, in the 1990s, I worked on a project called Harlem Live. The idea behind Harlem Live was simple. We would give inner-city kids who lived in Harlem video cameras and video-editing gear and teach them to shoot and edit and produce their own stories about their community. These were the days before Harlem had been gentrified, when Harlem was considered a very dangerous place to visit, let alone live. I have to admit that my first trip uptown was filled with trepidation. Would I get mugged? Would I get shot? Would I get killed? What in the world was I even thinking going up there? Didn't I watch the news?

This was also long before YouTube and before everyone had a video camera on their phone and posted their own videos daily, if not hourly, online on social media. In those days, there was no social media. The only video you could see was what was on TV. And TV stations had the only video cameras, and they decided what they would broadcast.

These kids all lived in Harlem, and at that time, the only things that they ever saw about Harlem on TV were stories about murders or robberies or rapes or fires. And they all watched a lot of TV. That was all I ever saw as well. That was all anyone ever saw. My "knowledge" of Harlem was a function of what I saw on the local news every night, but so was theirs. These kids were constantly inundated, night after night, year after year, with an endless stream of stories and pictures of violence in their own community.

This had an impact. It had an impact on them, it had an impact on me, and it had an impact on everyone who watched TV both in New York and anywhere in the world. It also had an impact on the people who made movies or TV shows in Hollywood. Endless episodes of *Law and Order* always portrayed Harlem as a violent place; it portrayed the people who lived there as violent, or pimps, or hookers, or drug dealers. The news stories, repeated ad nauseam, created a kind of reality in the mind of both the fictional TV writers and producers, but also in the minds of their audience. Thus, when they produced crime-based fiction, it would conform with the popular perception that had already been carefully cultivated. It was and is a vicious circle.

This is not so unusual; in fact, it is the norm. There is a deep connection between the way people and places are portrayed both in the news and in fiction. The fiction often drives the news choices—i.e., reports of shark attacks follow the movie *Jaws*—while fiction, to be believable, must cohere with the general perception of "reality" that journalism has already created.

Thus, the "news" stories that the local networks aired had a multiplier effect many thousands of times over in the rest of the media world. Say "Harlem" to someone, even in London or Tokyo, and it immediately conjured up the idea of a very dangerous, very deadly, drug- and crime-infested place. It told both the kids who lived there and the rest of the world at large that they lived in a very dangerous place, that their neighbors and the people on the street were probably not to be trusted, and that violence and crime were probably the norm. As Ed Murrow said in 1958, this instrument can teach. And so it could. It had been teaching both the residents of Harlem as well as the rest of the world what its vision of Harlem was. Repeat a lie often enough, magnify it through the incredibly powerful, persuasive, and addictive mediums of TV and film, and pretty soon, it becomes a reality that surpasses the truth. That is what had happened here.

As a result, you didn't see a whole lot of downtown New Yorkers taking the subway up to Harlem to check out the stores and restaurants. You didn't see a whole lot of tourists heading up to 125th Street to go shopping or just look around.

Those nightly broadcasts also had an impact on the kids. Over time, growing up in this kind of media environment, they came to believe that this was an accurate picture of their world. But the vast majority of the people who lived in Harlem did not commit crimes, rob people at gunpoint, shoot people, or burn their apartments down. Those things were, in fact, rarities. But they rated.

By giving the kids video cameras and teaching them to "make their own news," we were trying to introduce a very different perspective on what life in Harlem was really like. Far more important, we were trying to give the kids a sense of power over a medium that dominated their lives but that they had no control over. By giving the kids in Harlem both the video gear and the skills to use them, we were able to show them how to create TV about themselves that in a far more honest way reflected the world that they actually inhabited.

So they created stories about their friends, their schools, their families, their communities, and their lives. These were stories that were markedly devoid of crime, rape, murder, shootings, or fires. It was, in short, for them, real life, different from anything they, or everyone else for that matter, saw on television every night.

I cannot tell you what a liberating experience this was, both for the kids in the program and for us. What they captured in video was so vastly different from the stories that were on TV news every night, they might have been filmed in a completely different place. And, in fact, they were. They were filmed in the real Harlem, not the fantasy Harlem that existed in the minds of the news producers and reporters and filmmakers who all lived downtown.

For the kids, first producing and then watching those stories freed them from a kind of oppression by the media. You could see, as they worked on the stories and then screened them, a massive weight lifted from their shoulders. Until then, I had not really begun to understand the deep impact that the media has on people.

The problem in those pre-Internet days was that, having shot and produced so many stories, there was no place to show them to anyone else. No broadcast network, no local TV news operation, had any interest in what we were doing and what they had done. We were reduced to screenings in our offices. There certainly was no way to share them with either the larger community or the world at large. I even took the best of the pieces to NY1, a twenty-four-hour local TV news channel I had helped found in New York. "Here," I said, "are real stories about New York and the people who live there, made actually, not by your reporters, but by the people themselves. You have twenty-four hours to fill every day. Surely you can find a place for these."

The folks at NY1 were thankful for my contribution, but frankly uninterested in airing them—not even at 3:00 a.m. "This is not news," they told me. "These were not done by our reporters or our producers or our union camera crews."

Well, I suppose that that depends upon how one defines "news," but they certainly were not the kind of car crash that makes you slow down to take a look as you drive to grandma's house. There were no fires, no killings, no shootings, no rapes, no crime, no nothing—at least nothing that fit the old news adage "If it bleeds, it leads."

It didn't bleed; it didn't lead. It didn't even get on at 3:00 a.m. For the kids it was an eye-opening experience, but no one had their eyes opened more than I.

Harlem was being effectively financially and emotionally destroyed each time the news broadcast a scary story that got ratings. And, of course, Harlem was hardly alone. What was true for Harlem was true on an even larger scale for the rest of the world.

IT'S NOT JUST HARLEM

It is expensive for the BBC or NBC or CBS to send a film crew to a place like Sudan or Bangladesh or Iran to report on a story. It is complicated and involves massive logistical problems. And, for the most part, foreign stories don't rate. So the only time that any major media company will send a crew and report on a place like Sudan or Bangladesh is when there is a war or a major famine or an outbreak of Ebola, for example—you know, eyeball material, something that will rate. The company has to justify the costs.

In his book *The News About the News: American Journalism in Peril*, Leonard Downie Jr. quotes Tom Brokaw, who explained why *NBC Nightly News*, which he was then anchoring, never did stories about Africa. Said Brokaw, "There was no return on stories about Africa for the number one network in America." So the networks never bother to cover Africa, unless there is a mass killing or a deadly virus. Does this want to make you go to Congo for your next vacation? Does it make you want to invest in a new business venture in Cameroon? Close your eyes, and quickly state your perception of Africa. Disease ridden? Starving children? Civil wars? Atrocities? That is what you see because that is what you have been repeatedly shown. As Murrow said, this instrument can teach. And it does.

There is a kind of long-lasting destruction that comes from this kind of "journalism." It wrecks economies, it destroys the reputation of nations and cultures, and it destroys an individual's sense of self-worth and self-respect—that is, of an individual who happens to live in Harlem or Sudan or Pakistan. This is why you can have a president of the United States who calls for "a total and complete shutdown of Muslims entering the United States until our country's representatives can figure out what is going on."

Basing one's perception of the world, derivative only of what we see on TV, and Fox News in particular, reinforced by images in the movies, would seem a reasonable response, even if it had no grounding in absolute reality. It would seem to the general public watching the same news stories and the same movies that this makes a lot of sense. It seems the reasonable thing to do.

What local news in New York did to Harlem, international news and cable news did to Muslims worldwide, for example. And the perception of other people, places, countries, or cultures that is seeded and nurtured by the news media is soon reflected in the popular mass media of movies and entertainment. One would not believe a fictional film in which the archenemy was a Dane or a Norwegian. But make an action film with a Muslim or North Korean villain, and it is immediately believable. The fiction, repeated enough times, then becomes the reality. And the reality becomes the logical response and, ultimately, both policy and law, as well as innate truth. News begets fiction begets policy. It's a dangerous yet very powerful cycle.

About a dozen years ago, we were approached by the UNHCR, the United Nations High Commissioner for Refugees. This is the part of the UN that deals with more than 70.8 million refugees in the world today. The refugee crisis is an enormous problem of almost unthinkable magnitude. The global population of refugees is today greater than the population of France, yet these people have no home, no country, no jobs, no income, no means to support themselves, no access to education, no medical care, and often no food and water. It is almost incomprehensible, and the only people in the world who can or will deal with this is the UNHCR.

Yet, one of the greatest problems that the UNHCR faces is getting its story out to the public and the world so that it can garner the support it so desperately needs and, of course, so that it can change the general perception of refugees as dangerous. The UNHCR called us and said that it spent most of its time cajoling media companies like CNN or the BBC to please come to places like Darfur or Niger to do a story on the almost unfathomable refugee crisis that the world is facing right now. And, of course, with rising sectarian strife, civil wars, regional and religious intolerance, and climate change, the wave of refugees is only going to continue. Most of the time, its pleas to various major media companies went unanswered.

Most news organizations have neither the resources nor, frankly, the interest to mount the massive expedition it would take to send producers, crews, and reporters to places like Darfur or Mali. And what would they get out of it? Do their viewers even have an interest in stories about starving foreigners, most of whom have already been cast as a source of potential terrorism? How much of Brexit was driven by nightly scary news stories of armies of refugees streaming into Europe? How much of Trump's popularity is driven by nightly news stories of "caravans" of refugees streaming into America?

So the results of so many pleas to do a story were minimal at best. And even when, after a great deal of begging, CNN or CBS News would deign to send a crew and a reporter, all at great expense— the flights, the rental cars, the translators, the fixers, the hotels, the meals—they would arrive knowing nothing about the story they were to cover. How could they? Even the best and most dedicated journalist, parachuted down to a place like Darfur where they did not speak the language, know the culture, and or understand the complexities of the story, was lucky to be able to turn out a coherent one minute and twenty second story for the evening news, for this is all they were allowed.

And, of course, the financial pressures of the news business meant that they had a day or two, at best, to do their reporting. So the upshot was that on the very few and rare occasions that the network news reporter did arrive in Darfur or Mali, it was up to the UNHCR field workers, who actually lived there and dedicated their lives to these issues, who actually lived the story every day, to explain, as best they could, the story to the reporter and the producer and then hope for the best.

Having heard all this, it seemed to us that the most logical thing to do would be to cut out the middleman. That is, to cut out the news network itself—the producer, the camera crew, and the reporter. They were the weak link in the information chain, and the expensive one. They were, if you will, the medieval Church and monks of the operation—the old institutional power. But they could now be confronted with the digital printing press that each and every UNHCR field worker already had in their phones. How much more logical and reasonable, not to mention cost-effective, it would be instead to simply teach the UNHCR field workers to shoot, edit, and produce their own stories in the field and then to distribute them on the Internet for no cost to some 3.5 billion people?

And that is exactly what we did. Over a period of two years, we

put 165 field workers from the UNHCR through a series of vid-
eo-training bootcamps, run in both Geneva and Nairobi. In these,
we taught these field workers to shoot, edit, and produce video with a
smartphone, as well as the basics of how to tell a compelling story in
video and how to share it with the rest of the world on social media (as
well as offer a finished story to the likes of CNN or the BBC, which
they very much preferred, as it turned out).

The result was that we were able to empower the UNHCR to
take control of its own media footprint in the world. The organization was now able to use this incredibly powerful medium of video to
tell their story to the world the way it wanted to tell it, without the
hitherto all-important reporter, who knew nothing about the story
and could not have cared less, and without the destructive pressure of
having to produce some piece of sensationalism that rated. In effect,
the UNHCR was able to produce what we might call unmediated
media. It was an enormous success.

It was a success because there was now a direct media connection.
The people who actually lived a story, knew it, and cared deeply about
it were now able to tell it to the world in their own words. This is
what we call authentic authorship. This is what makes books powerful and believable, because the people who write them write from
both personal experience and personal passion. A book authored by a
production company would not be worth reading. There is too much
separation between experience and execution.

Reporters and production companies have no particular passion
for the subjects on which they report. They are paid journeymen who
simply have to do a job. One day, it's a mass shooting in Charlotte;
the next it's a farm story in Iowa. They don't answer to their own con-
science, they answer to the media company that pays them a salary,
and the interests of that media company are driven by commercial
revenue, which is driven by ratings. So to remove that power to produce the content, or rather to move it from the hands and the control
of the commercially driven media company to the hands of people
who have a real connection, interest, and passion for the subject about
which they are reporting is a fundamental change in the architecture
of how media works and what it delivers.

What we did for UNHCR was only possible because of the last gift of Steve Jobs. There was no need to purchase 165 professional video cameras and 165 professional editing suites, or to purchase a TV cable channel (last price, $1.2 billion) for UNHCR to free it to tell its own story in its own way. All the gear was already in the hands of the organization's people. All they needed was a bit of education in how to use the tools they already possessed.

And what UNHCR did, anyone can now do. What this means is that countries like Somalia or Bangladesh or Malawi can take control of their own media footprint in the world and change it. And this is not limited to NGOs or UN organizations, either—anyone can now do this. The basic construct of the media world has been fundamentally altered by a new invention. Anyone is free to embrace this and use it for the common good.

Let's take a look at the seemingly intractable conflict between the Israelis and Palestinians. The Palestinians, under the leadership of both the Palestinian Authority and Hamas, have often turned to violence, not in the expectation that firing a few rockets from Gaza into Israel was going to defeat the Israeli army, but rather that such acts would be a natural magnet to the world press. Violence, after all, rates. And the TV news networks, like rats to cheese, flock to Gaza or Israel proper to cover the same story over and over again and give the Palestinians what they think they want: attention.

While it works, it is a dangerous and ultimately futile way to get a story out to the world. And make no mistake, the Palestinians have a very good story to tell. I should know—in my youth I lived in Jabalia Camp in the Gaza Strip. There are approximately 2 million people who live in the 141 square miles that make up Gaza, and it is fair to say that living conditions are appalling.

According to a recent study by the Palestinian Central Bureau of Statistics, an astonishing 97 percent of Palestinian households have mobile phones. So let us say, just for argument's sake, that we were able to empower fifty thousand people in Gaza with fifty thousand smartphones to tell stories about what day-to-day life in Gaza was actually like. Now, suppose we were to continue to flood social media with those stories, day after day, week after week, essentially standing on a figurative rooftop yelling to the world: "Look! You! Poverty! Privation! Crowding! Hunger! And what have these people done to deserve this?"

Do you think that would, over time, have an impact? Instead of firing off rockets to hit schools and communities in Israel, suppose the Palestinians were to fire off story after story, vision after vision, of why they were so angry and so desperate and so in need of help. Do you think that would have a better and far more powerful impact on both world opinion and even on Israeli internal opinion? No rockets and no killing—there would just be stories. We all know the power of stories to teach. Use the medium to teach.

Gandhi drove the British out of India because he confronted them with a moral issue, not with violence. The power of the media, properly used (as opposed to commercially used), can be vastly more powerful than a gun. This, of course, has not happened in the Middle East, but all the tools are already in place to do this.

Let's take our UNHCR example and expand upon it even further in light of what this new technology could do with another kind of application.

Precipitated by the Syrian civil war, some 1.3 million refugees fled the Middle East and headed for Europe. This was a story that was covered over and over by the conventional media companies. It was a natural for ratings; it provided fantastic pictures as swarms of people headed west, often stopped at national borders, then allowed to pass through. It had all the makings of a great invasion movie—it was frightening, it was dangerous, and it was happening in front of the cameras—and of course it involved Muslims, lots of them, all swarming toward Europe. All the media news organizations had to do was to show up, which they did in droves.

The nightly news programs, and, as a consequence, the Internet, was soon awash in endless videos of veritable armies of migrants making their way westward, to God only knew where. The coverage had fantastic pictures but not a whole lot of depth. One thing that a viewer could notice was that pretty much every refugee had a phone. Most of the time, network news reporters would do stories of the migrants calling their relatives back in Aleppo or Damascus of Baghdad for that matter. If all of those refugees or a good number of them had smart-phones with them, then we can extrapolate that most of those refugees were probably not just calling home but also making videos about their voyage. After all, leaving your home and everything you have ever had or known and setting off on a one-way trip to save your life and your family has to be one of the most traumatic things you can ever do. So I think it is equally fair to extrapolate that most of those people were also shooting videos about what was actually happening to them in real time.

Now, if 1.3 million people were making videos on a daily basis, or even on a weekly basis, and those videos were about as accurate an accounting of what it was like to cross the Mediterranean in a small boat, to walk across Turkey and Hungary and Greece, to be held up in detention centers and at border crossings, how much of that video did you see on NBC or CBS or CNN or the BBC? Of all those millions of hours of real, honest, and direct accounting of a global tragedy unfolding in real time, done by the very people who were living it, how much of that raw and unmediated content did you actually get to see? Would the correct answer be none? And why would that be the case? Because the media companies that control what you get to see of the world did not want you to see it? Because they did not trust what was clearly there and so well documented? Because they had more faith in what some reporter who did not speak the language or understand the story in a way that an actual participant could? The answer to all of the above would be yes.

It is not that the content is not there; it is not that the potential is not there. It is not that the equipment to change the media equation is not there. It all is. It is that no one has known how to turn it on—until now.

TAKING CONTROL

What made the printing press so powerful a machine and so seminal in changing almost every aspect of society was that it was available to the individual. It was relatively cheap, easy to understand, and yet almost unthinkably powerful. Prior to the printing press, the control of information was in the hands of the Church and the monarchy.

The printing press took that control from them and democratized print.

The world of media that we created in the 1950s was, and to a large extent remains, in the hands of a very few corporations. They are the Church of the twentieth and twenty-first centuries. The iPhone and subsequent smartphones do for the world of electronic and digital media what the printing press did for the world of print—they democratize it. But that democratization can only happen when each individual begins to use his or her phone not as a place to watch but as a tool to create and share.

You may say, "Well, I am creating. I am taking photos of my cats and putting them on Instagram." That's fine; you get the concept. But you are not going to change the world with videos of your cats.

Carpe medium! Seize the medium. Use the same addictive electronic drug that has been used to manipulate your life and warp your world, but use it to change the world for the better. The machine is already in your hands. Take control.

11

What Has Oprah Got That You Don't?

OR
HOW TO START AND RUN YOUR OWN TV CHANNEL

BEYOND YOUTUBE

What has Oprah Winfrey got that you don't?

Well, for starters, she has $4 billion. On top of that, she has rich and powerful friends and the ability to bend public opinion, and she gets invited to all the best parties. She's also got a $42 million private jet; an estate in Montecito, California, that she bought for $52 million in 2001; 163 acres in Maui; a $14 million home in Telluride, Colorado; a 23-acre horse farm in California that she bought for $29 million; a 41-acre estate on Orcas Island in Washington state that she bought for $8.3 million; and 10 percent of Weight Watchers, which is now worth $400 million for her shares alone.

Now, Oprah was not born rich. She was, in fact, born as Orpah, after the character in the book of Ruth, but so many people mispronounced it that the Oprah name stuck. She was also born to an unmarried teenager, Vernita Lee, who worked as a housemaid. So how did Oprah do it? How in the world did she acquire so much wealth and power? What was her secret?

You can find the answer in a speech that the fictional TV anchorman Howard Beale gives in the movie *Network*, now a very successful Broadway show starring Bryan Cranston. Late at night, Beale hears a voice directing him to tell "the truth" to the world. Beale asks the voice, "Why me?" and the voice replies, "Because you're on television, dummy!"

And that is how Oprah amassed her fortune. Because she was on television.

When Oprah got into the television business, someone who wanted to be on television had to get a job with a local TV station, which is what she did. The station had the equipment, the broadcast frequency, the studio, the access.

Now, *you* are the TV station. You have the equipment, the studio, and the means of distribution, and it's all in your phone. But just making videos and throwing them up on YouTube or Facebook is not going to turn you into Oprah. In fact, for the most part, it's not going to make you a dime, or at least making money is going to be extremely difficult.

There are an estimated 1.8 billion users on YouTube. There's always a lot of chatter about YouTube millionaires, which of course, is possible, but rather difficult to attain. Recently, *Forbes* magazine, the same magazine that publishes the list of billionaires, also published the list of YouTube millionaires, those who make $1 million a year or more though their YouTube channels. There were twenty-five of them.

At the top of the list was PewDiePie, or actually Felix Arvid Ulf Kjellberg, a thirty-year-old Swede who has been putting content, which consists mostly of what is called "let's play" commentary on video games, on YouTube since 2010. His estimated revenue from his YouTube channel with more than five billion views is $12 million annually—not bad, but not exactly Oprah numbers.

Number twenty-five on the YouTube millionaires list was Lucas Speed Eichorn Watson, who goes by the name of SpeedyW03. He mostly comments on video games to his more than six million followers. He clears just north of $1 million a year in revenue—not bad for a twenty-eight-year-old who recently stated, "The earth is flat. NASA does not want you to see this picture." Draw your own conclusions.

Now, let's look at this objectively. Your odds on becoming a YouTube millionaire are an astonishing 0.00001388888 percent or about 1:100 million. The odds on wining the New York State lottery, in which you pick six numbers at random out of forty-nine choices, offers you far better odds, at 1:14 million. And playing the lottery requires a whole lot less time and effort than aspiring to be a YouTube star, which can mean working fourteen hours a day, seven days a week, creating and posting videos.

YouTube's system of paying its video content producers is both complex and opaque, but here are the basics: You must have at least one thousand subscribers to qualify for an ad-revenue share. You also need a minimum of four thousand hours of aggregate viewing time of your videos over a year. Already this is a lot more work than just filling in six circles on the lottery ticket. But if you want to go through all the effort and expense of making those videos, fine. Those aren't terrible barriers to overcome. If no one wants to see your stuff, you can't really expect to get paid.

If you are popular enough to pass that barrier, then you can start to look forward to a revenue share for advertising that gets appended to your videos. YouTube takes 45 percent of any ad revenue off the top, and of what remains, you can look forward to receiving anywhere from ten cents to ten dollars per one thousand "monetized" views. Only forty to sixty percent of views are worthy of being considered monetized views; that is, the revenue is variable based upon factors such as the location of the viewers and their "level of ad engagement," i.e., whether they click out or stick with it.

To cut to the bottom line, the top three percent of YouTube viewed channels, that is, those that receive one million to two million views per month, will typically earn $16,800 a year, based on a recent article in the *Financial Times*. And, according to the same piece, the top three percent scooped up 90 percent of all views.

To add to that rather bleak assessment, YouTube itself admits that most of the monetized channels make less than $100 a year, and ninety percent of those actually make less than $2.50 a month. You are not buying private jets or ranches in California on those kinds of numbers. You aren't even buying lunch.

The funny part here, or perhaps not so funny, is that YouTube's valuation, according to TheStreet.com, is somewhere north of $100 billion, and according to the same study, YouTube generates in excess of $15 billion a year in advertising revenue. On that, you can buy a private jet, which, of course, is what YouTube owners Larry Page and Sergey Brin did. They bought a Boeing 767, which costs $199.3 million, and that is before the interior decorating. Of course, they just added that plane to their existing Boeing 757, which you can pick up for $100 million. That only adds to their fleets of rather downscale Gulfstreams and apparently a German Dornier fighter plane as well— maybe to take on competitors like DuckDuckGo?

But this is the design of the content/compensation equation on the Internet, at least as it is currently constructed. You provide the content for no compensation, and the people who own the platform take the lion's share of the revenue that that content generates. It's a bad system, at least for the content creators.

Yet, there is a certain mythology, a certain mystique around the idea of being a YouTube vlogger or producer or owning a YouTube channel. According to a 2019 poll conducted by the LEGO Group, the toy manufacturer, along with the Harris Corporation, which does the polling, the number one career aspiration for children eight to twelve years old is YouTuber. The poll was conducted on the fiftieth anniversary of the first lunar landing, and LEGO was curious about how popular being an astronaut would be. Apparently it's nowhere near as popular as being a YouTuber.

Eleven percent of American kids wanted to fly to the moon, while a third wanted to video their lives and become professional YouTubers. Based on the study, LEGO may want to dial back on its new Mars Space Travel kits that precipitated the poll and perhaps would be better to offer kits that let you build TV studios out of LEGO blocks. Maybe Waylon Jennings should track a new song: "Mammas, Don't Let Your Babies Grow Up to Be YouTube Vloggers."

YouTube, of course, was first, but it is not the biggest. Let's look at Facebook, the killer of all content/advertising models. Fully 100 percent of the content of Facebook comes from users of the site who contribute their time and work and efforts to Facebook but receive no compensation for it. A great deal of the content is simply pho-tographs or videos or printed material, some of it original, a lot of it repurposed things that have appeared in other places that get shared on Facebook, again for no compensation. Facebook then runs ads against that content.

In 2018, Facebook brought in more than $50 billion in advertising revenue—that is, advertising against the attraction of the content for which they paid nothing. It's a great business model. It's like owning a supermarket in which farmers drop off their produce and chickens and lambs for free in exchange for "likes." The supermarket then gets to sell the stuff for money. In 2018, Facebook had a market cap of $468 billion. Its founder and principal shareholder, Mark Zuckerberg, had a net worth of $69.8 billion. That's a lot of chickens.

A lot of people say that you should launch your own channel on YouTube or Facebook. You will note that Oprah does not put her stuff on. That would be dumb. That would not be leveraging the potential value of content. You will note that Oprah has her own channel: OWN, the Oprah Winfrey Channel, which does put material on YouTube. OWN is owned (so to speak), or at least a good chunk of it is, by Discovery Networks. In 2017, Discovery paid $70 million for an outstanding 24.5 percent of OWN, giving Discovery a 70 percent ownership. That's how you afford private jets, not by putting your stuff on Facebook or YouTube!

Now, what does this mean for you? Since you are not going to kick the video addiction, you should at least make some real money out of it and maybe also take a measure of control over both the media world and your own life. Here's how.

WHAT DO YOU HAVE THAT DISCOVERY DOESN'T?

David Zaslav runs Discovery, Inc. In 2019, he was the highest-paid CEO in the United States, with a total compensation of $129.4 million. He must be doing a very good job of running Discovery. He is not getting paid that much money for generating millions of likes or followers. He is getting paid that much for generating real money from content.

Discovery was once a very small TV network, founded by John Hendricks, a former employee of the University of Maryland, who had an idea for this new cable thing. Cable had just arrived, putting lots of channels into people's homes. These new cable channels had lots of empty space and needed content. Hendricks had an idea for how to fill that space: documentaries. In 1982, he set up something he called the Cable Educational Network to provide documentaries to the new cable channels. Today, Discovery owns the Oprah Winfrey Network. It also owns Discovery, TLC, HGTV, the Food Network, the Cooking Channel, the Travel Channel, the DIY Channel, the Science Channel, Animal Planet, and more. In fact, it owns nineteen channels, and this does not begin to include all the international holdings and channels. It has a market cap of just over $21 billion.

It's another media-content income model, and it may be a bit of a better model for us than YouTube. Hendricks, seeing ahead of almost everyone else the potential that a new technology (in this case cable) offered for a new way of delivering content, is also a good role model for us to follow. What cable was to Hendricks in 1982, online nonlinear digital can be for us today. And let us remember that Hendricks was offering documentaries, that is, video with intelligent content.

As newspapers do with print, Discovery also puts content up on cable channels and then charges advertisers to piggyback on the audiences that it attracts. On the bigger Discovery channels, a single thirty-second ad spot can cost an advertiser as much as $100,000. In 2018, Discovery Communications brought in $13.8 billion in revenue.

The problem that Discovery has is that it is the product of a 1980s technology: cable. Its content is linear. That is, it has to continually air and thus endlessly produce new content because it goes out in real time. As such, there is a program on at 8:00 p.m. and then another one at 8:30 and another one at 9:00 p.m. and so on, twenty-four hours a day, 365 days a year, across nineteen channels. The machine eats content. After all, there is no point in owning a cable channel and not putting something on it. So Discovery has a voracious appetite for content. It essentially has to fill 55,480 hours of content a year, just for its US holdings. That's a lot of TV shows.[32]

But Discovery (and every other cable network) was architected

---

32 There is no point in owning a cable channel unless you are running programming. Programming is the lifeblood of cable. No shows, no ads, no ads, no content. If each of Discovery's nineteen channels runs eight hours of original programming a day, then each year it must produce 55,480 hours of content.

before there was an Internet. In those days, there was only cable, and cable was relatively new. It was a good idea, based on what the distributive technology was at that time. In those antediluvian days of early cable, there was a strong and solid barrier to entry for any competitor. You might sit at home and make the most amazing TV shows in your living room, but unless you could put them into people's homes so that they could see them, and hence charge advertisers to talk to the audience you had been able to aggregate, they were essentially worthless: no viewers, no advertisers, no revenue.

You, on the other hand, are nonlinear. Netflix is also nonlinear, as is Apple TV, Amazon, AT&T, and every company getting into the next generation of video-driven content. Cable was old. Nonlinear is just getting started. Here's your chance to do what John Hendricks did in 1982—get in early.

WHAT IS A CHANNEL?

The term "channel" is really a remnant of another time. What we are really talking about here is a platform upon which to post your video so that you can generate revenue from it, use it to influence people and public opinion, sell your services or wares, or all of the above. Because your cost of production is so low (iPhone) and your cost of distribution is so low (free), you need no longer pander to the lowest common denominator to accrue viewership. You can be profitable on a much smaller scale, and, as such, you can use this incredibly powerful medium to do more than simply sell Cialis.

In your phone, you have all the tools to create compelling video. (We'll discuss just how to create compelling video in the next chapter.) But having compelling video that no one ever sees is a waste of time and effort. And taking all the time and effort to create compelling video and not getting paid for it is also a waste of time and effort. A new opportunity is before you, no matter what business you are in, no matter what your ultimate goal is, so take advantage of it. Here's how.

As you can see, turning your hard-made content over to Facebook or Instagram or YouTube is essentially giving your work away for free so that someone else can profit from it. That is nuts. Better to create your own platform, your own website, your own "channel," and put your content there. Creating a platform or a website or a destination online for your video is today relatively simple to do. There is so much software that is today plug and play. Anyone can do this.

My own personal preference for simple website builder is Wordpress. But there are many others that you can choose from. Here are a few of the best, but they are all pretty much the same:

- Wix
- Duda
- HostGator
- Simvoly
- Squarespace
- GoDaddy
- PageCloud
- Strikingly
- uKit
- Weebly

Not so long ago, building a website, particularly one that was capable of broadcasting video (and taking payments), was incredibly complicated. You had to know HTML, you had to know how to code, you had to be something of a computer-programing genius.

In 1994, in the early days of the web, a company called Razorfish was founded by childhood friends Craig Kanarick and Jeff Dachis. It was run out of their apartment in what was then the fairly dangerous and edgy Alphabet City in downtown Manhattan. Kanarick and Dachis were among the first to learn HTML, the coding that allowed the creation of websites. This gave them an edge. They could build websites before anyone else even knew what they were talking about.

Their first project was a website for the New York Botanical Society. For this, they were paid $20,000. Razorfish generated over $300,000 in revenues in 1995 and over $1.2 million in 1996. These were the early days of the web. Six short years later, the company would have a valuation of over a billion dollars. The two founders, Kanarick and Dachis, were both interviewed by Bob Simon for *60*

Minutes. When he asked them what, exactly, they did for a living, neither could really answer the question. The interview was generally believed to be a disaster. But what they did was make websites, before anyone else knew how to do it. That was once worth a fortune. Today, using any of the companies listed above, and many others, you and any nine-year-old, can make a website that is far more functional, powerful, and sophisticated than anything Kanarick and Dachis could have dreamed up, and you can do it for next to no cost or even no cost at all.

The reason you want a website of your own is that you want a home for your content—a destination that has an identity, a place that is singularly yours. You want to own it and control it. This is the digital equivalent of a store. Putting your video content on YouTube or Facebook is the digital equivalent of renting a table at a flea market and having to give up forty-five percent of everything you sold for the privilege of renting the table. It has no identity. You are a visitor in someone else's home. With your own site you have your own home. And this holds true, whether you are using your newly acquired video literacy to move a political needle, to promote your business or your cause, or to make a buck. The power of video is extraordinary, and the most important thing here is that you control it so that it no longer can control you.

So far, the only workable revenue model for a content business, online or otherwise, has been to sell advertising against the eyeballs aggregated. This may still work, but I think that the Internet allows entirely new ways of monetizing your work.

MAKING MONEY IS NOT A CRIME

People are always confusing news and journalism. News is something that happens. News can also be information. Journalism is a business that is specifically designed to turn news, which is free, into revenue. The primary purpose of journalism is to make money, because, all things considered, if there is no revenue, then there is ultimately no journalism.

I taught at both Columbia University's Graduate School of Journalism and NYU's journalism school for many years, and most of the people who are attracted to being journalists do so because they have been convinced that they are embarking on some kind of holy mission to uncover the truth and alter the world. Their heart, at least, is in the right place. They are, in fact, training to be low-paid sharecroppers on the content plantations of people like Rupert Murdoch. Since Murdoch's business model does not involve running ads against content that he has wheedled people into giving him for free, he has to do the next best thing, and that is get content made for him at the lowest possible cost. That's where the journalist comes in: hard working, dedicated to a cause, incredibly earnest, and dirt poor.

This image is constantly reinforced in movies and TV shows. The journalist is always portrayed as some poor schmuck living in a crap apartment, driving a crap car, wearing crap clothes, and eating crap meals but unrelentingly in pursuit of the noble truth—rumpled, bad hair, crap salary, but by God, driven to do good.

There is a famous quote that floats around the world of journalism, and particularly around journalism schools. It comes from the fictional nineteenth-century Irish bartender, Mr. Dooley, who was created by *Chicago Evening Post* writer Finley Peter Dunne in 1893. Mr. Dooley's most lasting quote was, "The job of the newspaper is to comfort the afflicted and afflict the comfortable."

This is, of course, pure nonsense. The job of the newspaper is to make money. If the newspaper is not making money, then pretty soon there is no newspaper. This is true not just for newspapers, but for any media business.

But, as with cops who fire off thirty rounds at the "bad guy" and become heroes because they have seen it on TV and the movies, this identity for the journalist has become fixed in the public mind, at least since Woodward and Bernstein took down a president. What makes the money in the media business is the advertising. The content is just the stuff that gets placed between the ads.

Zuckerberg and his pals took the genius step of no longer even having to pay for the content. Like Tom Sawyer conning the neighborhood kids to whitewash Aunt Polly's fence, the world of social media has convinced billions of people, literally, to become unpaid content contributors with their phones.

Today, everyone is a journalist, one way or the other, since most of us are content creators and providers. Even if you only post your family photos on Instagram, you are a journalist—admittedly with a fairly limited patch. And, since we are now mostly all journalists, it is time to start making some real money out of the media business that we all support so strongly but don't get a lot back from. Carpe medium.

Now, let's talk about video, that most powerful and most addictive medium, and a relatively new invention. When television first started in the 1950s, the content was pushed through the air. As a consequence of that, there was a physical limit to the number of channels that you could have. It takes a lot of space on the electromagnetic spectrum to transmit analog pictures and sound. So in the United States there were three networks: ABC, NBC, and CBS. In the United Kingdom and most other European countries, there was but one. And since there were so few networks, the networks had to provide all things to all viewers: news, sports, entertainment, comedy, game shows, weather, politics—everything. When cable came along in the 1980s, it expanded the number of possible channels. But the habit of having one channel show all things was deeply entrenched in our understanding of what a channel was supposed to do. So the early cable channel operators, having only their experience of having watched broadcast networks, aped the broadcast channels, and also tried to offer all things all the time.

For some time, I worked as a video journalist for the Christian Science Monitor, which had a TV news show (and later a channel itself). *The World Monitor*, which was its nightly hour-long news program hosted by John Hart, appeared on the Discovery Channel. Discovery in those early days thought it should do news, just like CBS or NBC. But pretty soon, the cable networks discovered that it did far better if they specialized and owned a niche—hence the Travel Channel, the Food Network, ESPN for sports, HGTV for real estate, and so on. This proved to be a very successful idea.

Now, as we move from linear cable to nonlinear online digital networks (i.e., your website), we can take away an important lesson here. Specialization works.

If you think you are going to build a content-agnostic platform like Facebook or Instagram and try and be a place for all content for all people, you can stop right now. That is not going to happen. In fact, I would guess that broad-based platforms like Facebook and Instagram are probably going to fractionalize in the future and specialize, much as cable once did.

But for you, there is lots of opportunity in even more fine-grained specialization than just "food" or "travel." The next time you are at an airport waiting for a flight, take a look at the magazine store. See how many different magazines there are, aside from the predictable ones: *Salt Water Sportsman, Dollhouse Miniatures, Bee Culture.* There are, in fact, 7,176 different magazine titles in the United States alone, and this is down from an all-time high of 7,390 in 2012, so the magazine business has not suffered much from the arrival of the Internet.

You can see the reason for that if you take a look at the masthead of any magazine. The staff is kept small, so the overhead is small. The revenue comes from advertising and subscriptions. The key to profitability is to have the cost of manufacturing content lower than the revenue that that content derives. We are going to follow exactly the same model for your online digital channel.

Because you are going to make the video content, video being so much more addictive than print, on your phone, your cost of manufacturing content will be low— actually next to nothing, nothing but your time. If you can generate more revenue through advertising, subscription, or online transactions (or a combination of all three), congratulations! You now own a successful channel all your own. See how easy that was?

GO WEST, YOUNG MAN!

On January 24, 1848, gold was found by James W. Marshall at Sutter's Mill in Coloma, California. This discovery precipitated the California gold rush and the great movement to the West Coast. New England newspaper publisher Horace Greeley captured the spirit of the times when he made famous the phrase, "Go west, young man!" That was where the future was, in the west. It was new and open. Today, the gold rush is in online digital channels that you can build and own on your own. It is also new and open.

Between 1850 and 1880, when American settlers started heading west for California, they massed in wagon trains in Independence, Missouri, before starting the long and arduous overland journey. They did not set out blindly but followed established trails across the American prairies and deserts that others had trodden before them. These were the now-famous trails such as the Santa Fe Trail, the Chisholm Trail, the Mormon Trail, and so on.

Those who went before them had carved great ruts in the ground, the wheels of the covered wagons being covered with iron, which dug into the soil. Over time, the ruts became the equivalent of interstate highways, all headed in one direction—west.

When the wagon trains started on their journey to California and other points west, the wagon train masters would say to the assembled just prior to departure, "Pick your rut. You're going to be in it for the next two thousand miles."

If you are with me so far, then you are like those brave pioneers, headed out to seek your fortune and your freedom in the digital wild west. So, as those wagon train masters once said, pick your rut; you're going to be in it for a long time. That is, when it comes to launching your own channel, pick a topic that you have a real passion for and also an understanding of, and stick with it. You are also going to be in that rut until you get to the gold at Sutter's Mill.

The best topic for a digital channel of your own is a business that you are already in or a topic that you have a real passion for and know well. Hence, the digital channel becomes a vast multiplier of what you already do. For example, if you are in the real estate business, you might already have video on your website for selling homes or offering rental properties. I have seen thousands of these, as have you.

But take a look at what passes for video on a typical realty site. It's generally no more than a series of still photos or the video equivalent, sometimes set to music, sometimes not, with a set of Ken Burns movies attached to them. Now, in all honesty, if you saw that very video on HGTV as a TV show, how long would you watch it? A minute, maybe, but only if you had something else to do at the time.

That kind of video, appended to selling homes, is living death. And it screams to the viewer of your site, "I am a bore." The people who are on your site looking for a home are also the same people who are watching *House Hunters* on HGTV. Now, if you had a lot of money, you might buy a thirty-second spot on HGTV and run an ad for your realty company. You might do that, but at $100,000 per ad, that is not going to happen.

Here is something else you can do instead. You can shoot, on your phone, and create, on your own and at no cost, your own version of *House Hunters*, but visiting three houses that you want to sell. You can even be the star of the show on your channel. Do you see how this works?

Now, according to Nielsen, twenty-five million people watch *House Hunters* every month. There is a reason for that. It is fun to watch. It has all the elements of a great and compelling (and addictive) TV show. Now, here is the really weird part of how this currently works. You get sucked in to watching *House Hunters* or *House Hunters International*. You can spend an hour or two looking at houses, being shown the kitchens, the living rooms, the two-car garages, the gardens. You are effectively being taken on a tour of a house that is for sale by a real estate agent—you and 25 million other people. You may, in fact, fall in love with one or two of the homes that you are looking at. But here's the weird part. Having been shown all these houses that are for sale, you can't buy any of them. It's like spending an hour walking around Bloomingdales, falling in love with some shirt or shoes, and then being told, "Sorry, not for you."

In the case of *House Hunters*, it is not selling houses. Rather, it is selling the viewer, who is drawn and addicted to the content, to the advertiser. Why not learn from HGTV and *House Hunters* and Discovery and instead use that same medium, that same format, that same addiction to sell *your* houses?

When it comes to monetizing your channel, it is far easier if you build a channel that is a multiplier for an already-existing business. As I mentioned in an earlier chapter, there are thirty million small businesses in the United States, so there is room here for thirty million channels that can be a multiplier for an already-existing revenue source. Why give the money to YouTube when you can own it yourself?

Let us say that you own a restaurant. You can see how many food-related TV shows there are already. It's a very popular venue because people have to eat, and they do that a lot. Go to any restaurant's website however, and what do you see? A menu and some photographs—very 1980s. If there is video, it is like the video on the realty sites—deadly boring.

But take a look at what works on the Food Network, for example. An astonishing eight in ten adults watch some kind of food show. Yet, what is really crazy is that not one of those food shows is related to the idea of actually filling seats in a restaurant or selling food. Oh, it may happen tangentially. A chef may become famous by appearing on some kind of food contest, and as a consequence of that, people will come to seek him out. But this is a very roundabout way to use the medium to drive business, and even here, the lion's share of the profits from the work of the chef are going to the network.

Do you see the potential here? If you have a realty company or a bakery or a restaurant, or if you make granola, you can see how this can work. If you don't have a product to sell as part of a whole media package, then find someone who does and partner with that person. The technology of online transactions and immediate gratification is going to render advertising archaic, but a whole new revenue stream is just being born, once again, thanks to the technology. It makes far more sense to marry the addictive content that video is, no matter what platform it appears upon, to a transactional purpose. It is more direct, and it is more honest. And you can do this. And you can do this now.

WHAT NAPOLEON TAUGHT US ABOUT MARKETING

In 1797, France was on the ropes. The French Revolution of 1789, which had begun with such high ideals, was degenerating into the Reign of Terror, as the Revolution and the revolutionaries began to consume themselves at the guillotine.

From the ranks of the French Army arose a hero who would not just lead France to victory after victory but also would eventually become one of the most written-about characters in history:

Napoleon. Starting in the late 1790s, Napoleon began to win battle after battle, ultimately taking France from the edge of invasion and defeat to the edge of the complete conquest of almost all of Europe in only a few years. It was an astonishing achievement.

Now, Napoleon was a great general, but there were many great generals in Europe at that time. However, Napoleon did something truly revolutionary that helped his conquest of Europe a great deal, and it is an interesting lesson for us and our use of video, social media, and networks today.

Battles in Europe at that time were fought by small, highly trained, and highly professional armies of around thirty thousand men. Soldiers were this way because they were, while quite good at what they did, also quite expensive. They were professional. This is how everyone went to battle. You have, no doubt, seen images of this in movies or TV shows or books—highly polished uniforms in bright garish colors, sabers, horses and so on.

Napoleon, on the other hand, introduced the concept of the Grande Armée. He drafted every male he could find and put them through some very basic training, then off to battle. At its height, Napoleon's Grande Armée numbered 685,000 men. Well, when you show up with and army of 685,000 men on the field of battle, it's a pretty good chance that you are going to win. And win he did, over and over and over.

Now, what does this have to do with video, social media, having your own channel, and making money? A lot, as it happens. Everyone understands the importance of using social media, which is just another platform, to promote their business. While we don't want to give our content and our revenue to YouTube or Instagram, we are certainly in favor of using YouTube and Instagram and Facebook and everyone else to drive traffic to our own site. But most businesses send one or two or perhaps even a dozen social-media warriors onto the Instagram field of battle. Small companies DIY it; larger ones hire PR or ad agencies to help them construct their postings on Instagram or Twitter at a very large fee. These are the "professional soldiers" of the digital wars—highly trained, very polished, very expensive.

Now, let's take a look at an average company, say Bed Bath & Beyond, for example. It has 1,550 stores and sixty-five thousand employees. Now, Bed Bath & Beyond, as it happens, does not buy a lot of ads on HGTV, a Discovery channel totally dedicated to home improvement and decorating, just the kind of stuff that Wayfair, who does buy a lot of ads on HGTV, sells. So you can see why Wayfair would be willing to spend tens of millions to buy ad spots to get their product and their company in front of the viewers that HGTV is attracting. Their viewers are Wayfair's buyers.

However, when TV was invented, it didn't come with an instruction manual, so let's think a bit out of the box. Let's think like Napoleon for BB&B. Going to the BB&B website you'll see there is not a whole lot of video, but there is a lot of stuff to buy. Now, Wayfair spends all those millions on TV ads in the hope that if you watch the ads on HGTV, it will convince you to either get in your car and drive to the Wayfair store near you, or to go online and go to the Wayfair site and buy stuff. This is a bit complex.

Now, suppose that BB&B were to create a BB&B channel on its website that had the same kind of compelling home-improvement shows that HGTV, for example, already produces and airs. Those shows are pretty popular. *Fixer Upper*, a home-improvement show starring Chip and Joanna Gaines, is HGTV's most popular show.

So let's suppose that in a brilliant counterprograming move, BB&B has its own channel, the BB&B Channel, and on that channel, which can only be seen on BB&B's website, it airs shows with compelling characters who also fix up houses, but they use BB&B products to do it. Does that make sense? Does that connect the content with the revenue more directly? Because now when you are watching the BB&B *Fix-up Show*, you can also click and buy right there—no need to move a thing, except your finger.

Every one of the sixty-five thousand employees at BB&B has a smartphone. I am also willing to bet that all of those employees at BB&B are also posting their own stuff on their own sites on Instagram or Facebook or Snapchat or Twitter. Now, let's for a moment suspend disbelief, and let's say that in the digital world, it is not unreasonable to ask your employees to post, say twice a week, things about their job and their company. Who, after all, knows BB&B products better than the people who work there? Who has a better interest in helping grow the company that is paying for their mortgage and sending their kids to college? They are happy to explain stuff to you in the store, so why not online? Now, we have sixty-five thousand BB&B employees posting a small video or text or photo to social media twice a week, all about the great stuff that BB&B has and does, with the object of sending people to the BB&B Channel.

Our sixty-five thousand employees, armed with their smartphones, are like Napoleon's 650,000 soldiers taking the field—a Grande Armée of digital content creators.

Let us say, just for argument's sake, that each of the BB&B employees has one hundred followers. I think this is low, but let's use it just for the case. So now, our social-media messages about BB&B and its great products and services are going out to 6.5 million people twice a week. And, let us now postulate, again, just for argument's sake, that 10 percent of those followers then share or repost the messages that they have gotten from their friends, and each of them has one hundred followers. So now, our BB&B message is going out, twice a week, to an astonishing 71.5 million people—twice a week, and at no cost.

Just for fun, let's put those numbers in perspective. The Oprah Winfrey Network just had its best quarter ever. Do you know how many people were watching OWN in prime time? 599,000 total viewers.

And what is the purpose of all of this social media? To drive people to the BB&B site where they can see the show and buy things that the show is showing them, so to speak. And what BB&B can do, you can do. All you have to do is, as Nike says, "just do it."

And, here's a thought for Nike, relative to the power of this new medium. Nike buys lots of ads on TV. Last year, Nike spent an astonishing $3.5 billion on advertising. That is a lot to convince people to go to the store and buy sneakers.

Nike buys lots of advertising spots on, among other things, the World Series. People who watch the World Series probably buy lots of the sporting equipment that Nike makes and sells. The World Series

TV rights are owned by Fox. Bloomberg reported that Major League Baseball sold the TV rights to the World Series to Fox, through 2028, for $5.1 billion. Fox makes its money back by selling advertising on the World Series to companies like Nike.

But suppose that Nike bought the rights to the World Series after 2028. They could afford to do it. Fox has a market cap of $20.99 billion. But Nike has a market cap of $109.3 billion.

Now, in our theoretical model, suppose, as Nike now owns the broadcast rights to the World Series, that the only place you can watch the World Series in 2029 is on Nike.com. Then everyone who wants to watch the World Series will have to go to Nike.com to see it. One of the great advantages of watching it on Nike.com is that it is ad free. That is because, instead of running ads, Nike has surrounded the broadcast with click-and-buy ads for Nike products. So you can buy Nike stuff all day long as you watch the games.

Get the concept? It is content married to transactions.

BACK TO HARLEM LIVE

That's all fine if you own a restaurant or a real estate company or the World Series. But what if you just want to change the world? What can you do then?

Last year, we had an idea. We thought it would be interesting to try to resurrect the Harlem Live idea. Now that everyone has smartphones, the gear is all there. And now that there is an Internet, the distribution system is all there.

So we got in touch with our old friend Samson Styles and shared our idea with him. He took us to Brownsville and East New York, two neighborhoods in Queens that have not been gentrified. The only time local news goes to Brownsville and East New York is when there is a shooting or a robbery. It was just what we were looking for.

Through Samson, we met Andre Mitchell, who runs an organization called Man Up. We explained our idea, and he set up a community meeting at the local junior high school; 150 people attended.

We asked the assembled crowd to put up their hands if they felt that local news showed their community in a bad light, and 150 hands went up. We then asked them if they would like to change the kind of local news that was being broadcast nightly. Again, 150 hands went up. Finally we asked who had a smartphone. You guessed it—150 hands went up again.

So there are 150 broadcast-quality cameras and 150 edit suites in this room. That makes Man Up TV the largest TV station in New York City right now. For the next three months, we trained those volunteers to shoot and report and edit on stories in their community. As Napoleon knew, training is essential. But the real problem is revenue, because if there is no revenue there is no journalism.

Now, here's the interesting idea. Just as each of those phones can make a news story, they can also make a TV commercial. So we simply commissioned people to go out and make ads for local merchants for the new "station." Charge the merchants $100 for the commercial—half to the station, half to the person who had the phone and made the ad. It's an entirely new model, but it's an entirely new world. And it's a model that can be applied anywhere in the world: journalism of the people, by the people, and for the people.

COMING UP NEXT!

The experiments in Harlem and Brownsville and with the UN are but a tiny microcosm of what is possible. It was one thing when Martin Luther seized control of the printing press, but the real change would only come when millions of people started writing and publishing.

In the next chapter, we are going to take a look at the enormous and as yet largely untapped potential that the unique combination of smartphone and Internet presents. It's going to be a whole new world, for those who want to be a part of it.

12

This Instrument Really *Can* Teach, If You Let It

DEATH TO AMERICA

On November 4, 1979, Iranian students loyal to Ayatollah Khomeini, who had initiated the Iranian Revolution earlier that year, entered the grounds of the US Embassy in Tehran and seized control of it, taking hostage fifty-two American diplomats and citizens. This came to be called the Iranian Hostage Crisis.

These are the kinds of events that news organizations live for. They drive ratings. Of course, every American news network was all over the story, but news, being a competitive business, is always looking for an edge. Just two years prior, ABC, the third-rated of America's three TV networks, had put Roone Arledge in charge of ABC News. Arledge was an unusual choice for a news division. He had been the unquestionable star of ABC's sports division. Also coming in to run the third-rate network's failing sports department, he had, in only a few years, taken if from unknown to first place and made it a money-making machine for the network with inventions such as *Wide World of Sports*.

Now, the network was hoping that Arledge could work his television magic on their anemic news division. The Iranian Hostage Crisis gave Arledge the opportunity he had been looking for.

On November 8, just four days after the seizure of the embassy,

Arledge decided to start running a nightly news show devoted entirely to the hostage crisis. It was called *The Iran Crisis: America Held Hostage*, and he felt this would be a ratings winner against NBC's *Tonight Show*, the industry leader. *The Iran Crisis: America Held Hostage* was originally hosted by ABC News correspondent Frank Reynolds but was shortly taken over by State Department reporter Ted Koppel. A few days after it started, a producer on the show got the idea of tagging it by the number of days that the hostages had been held—Day 4; Day 15, Day 34 and so on. It was a nice dramatic touch.

No one really knew how long the special series would run. It would, in fact, run until Day 444, more than a year later, when the hostages were released. So, ABC News ran 444 half hours (initially twenty minutes) devoted entirely to one subject—the Iranian hostages.

Now, you only need sixty classroom hours at Clemson University to qualify for a PhD in Bioengineering. ABC News was providing to the nation, at no charge, 222 hours of classroom instruction on just one topic, or nearly four times what is required at Clemson for a doctorate. Yet, if I were to strap you to a chair and make you watch all 444 *America Held Hostage* programs, your brains would probably drip out of your ear. That's tragic, because with 222 classroom hours on one subject, everyone in America should, by the end of the crisis, have not only been fluent in Farsi but also been rather well educated in the whole Islamic fundamentalist revolution that was happening in Iran. This kind of education might have come in handy a few years later when President Bush suggested invading Iraq and instead set the Middle East on fire for a generation or more. An educated public, conversant in the history of both Islam and the Middle East, might have been a good deal more resistant to the idea of invading Iraq, and a massive tragedy could have been averted.

This notion of using the media to educate people, with respect to news at least, need not be applied only to crises. It could be applied universally. In a 2018 study by Pew and *USA Today*, it was noted that 47 percent of people still go to television as their primary source of news.[33] That is still a substantial percentage of the population. And, one may assume that it is the same people who go there night after night, a dependable return audience. If that is the case, and I think it is, then there is an enormous opportunity here that is not being exploited. Each night, the news programs begin from scratch. Three or four stories, all running 1:20 or so in length, and most disconnected from the rest. But, as you have the same viewers night after night, week after week, year after year, in fact, all tuning in at the same time to the same location for the same information, you could do something both interesting and vastly different. You could create a national curriculum. In other words, if you knew that you were talking to the same people, night after night, which you in fact are, you could create a curriculum on specific topics that built upon itself over time.

You could educate a nation on topics like climate change or the economy or Islamic fundamentalism, slowly, brick by brick, over months and even years. There is no reason to start from zero every night and then again back to zero the following night.

This very much takes us back to our prior conversation about arc of story. We are a species that relates innately to storytelling, and arc of story is in a strange way directly innate to news, if you view news as "the first draft of history." What, after all, is history but a compendium of stories, mostly of people—Alexander the Great, Caesar, Napoleon, George Washington—and what they did, their arc of story. Stories, constructed in this way, of course, are the way that we as a species educated our succeeding generations for thousands of years. So here we have this magnificent, and addictive, no less, machine that is specifically designed to crank out stories.

Instead of scatter-shot car crashes and vaccine mysteries, let us look at news as history in process, with all the best elements of history: character plus arc of story.

Part of the problem here is that we inherently look at education

33 Mike Snider, "Americans Prefer to Watch News, on TV and Online, Rather Than Read It, Study Finds," *USA Today*, December 3, 2018, https://www.usatoday.com/story/money/media/2018/12/03/most-americans-prefer-watch-news-not-read-study-says/2189310002/.

as boring. The irony, of course, is that we all go home from school to get our real education on TV and YouTube, with news, movies, TV shows, and all else. The problem with the TV/YouTube education is that it teaches us things like fear of Muslim terrorists or fear of kidnappers or fear of plane crashes or serial killers in your backyard.

This notion of marrying the educational power of storytelling to news is not limited solely to Islamic fundamentalists or global warming, although it would work for them. News as done on television is like Alzheimer's. Each day it forgets what it dealt with yesterday and who it met on the tube. It starts with a blank slate each day. This is a mistake, and it does not leverage off one of the most basic rules of storytelling—characters you can relate to and sympathize with.

Looking at news, let's take a theoretical example. Let us say that you work for a local TV news station in Los Angeles. Let's also say that the Olympics are coming up in a year in, say, Tokyo. Now, in a city like LA, you are going to have a lot of Olympic hopefuls who are going to be working really hard to make the team. This is a classic local news story. In 1988, I covered Olympic hopefuls for CBS's *Sunday Morning*. It's an easy story to do. Character plus arc of story would mean, are they going or not?

But let's take this very simple and very repeated story and do it a bit differently, at least in theory. Let's say we focus on the Special Olympics. Let's say that we find a Special Olympics athlete who is living in LA and working hard to make the cut. But let's say that instead of just doing one story one time, we instead say that we address this Special Olympics contender not as a one-off but as a character in an ongoing "reality show" that also happens to be on the news. That is, we keep coming back to her, and her family, and her coach, and her competitors, and their families during the next eighteen months or however long it takes before the winners all decamp for Tokyo. We get to know them as characters that we care about, the same way we get emotionally bonded to characters on a program like the *Real Housewives of Wherever*. But these are the *Real Special Olympics Athletes*.

Then, let's expand the concept. Let's say that we covered a number of these para-athletes, and we got to know not only them but also their coaches, their families, their spouses, and so on. Now, in the course of the next eighteen months, all of our characters are going to have interesting events in their own lives, but their lives are all going to intersect, that being the nature of competitive sports. Are you starting to see the show here, as opposed to the news?

Now, an ongoing reality TV series is, I think, a vehicle both to entertain and to educate. Because as we get invested in our characters, their arc of story, we are now interested enough to let the story teach us. And that is what stories are very good at—teaching. And this story, this ongoing reality TV series about real people and real lives, also becomes a vast educational tool to learn about not just Paralympics, which is pretty good, but also how people deal with their own handicaps and how government can help or hinder. When you care about the characters, you care about the things that they care about.

THE ARCHERS

You will recall that when a new technology arrives, it does not come with an instruction manual. Freed from the commercial constraints of having to maximize the audience to sell adds, the BBC, which is noncommercial, engaged in a very interesting experiment in character/arc of story media for the purpose of education, with some astonishing results. And it did it more than sixty years ago.

On Monday, May 29, 1950, the BBC began broadcasting a radio drama entitled *The Archers*. It was the story of three farm families: the Archers, Walter Gabriel and family, and George Fairbrother. The three families represented three different kinds of farming. The Archers farmed efficiently with little cash, the Gabriels farmed inefficiently with little cash, and George Fairbrother was a wealthy businessman farming at a loss for tax purposes.

The radio program had originally been designed as an educational program for BBC listeners in the countryside. It was hoped that in the postwar years, radio could be used to teach farmers newer and better farming methods, wrapped in the guise of a radio drama. Thus, the Archers were the model and the Gabriels, the failing past.

Today, *The Archers*, still on BBC Radio 4, remains one of not just the BBC's but also Britain's most popular dramas. Having aired over

19,100 episodes, it is the world's longest-running radio drama. And it is still educating, and certainly entertaining.

What made and continues to make *The Archers* work is once again that classic combination of characters plus arc of story. Of course, over the course of nearly seventy years, there have been many interwoven arcs of story, but the basic premise still remains. As well, new characters have appeared on scene, and a few, but not many, have died off—often to the mourning of all of Britain. The attachment of the public to *The Archers* and its fictional characters is a great lesson for us in the power of character and storytelling. The fact that *The Archers'* primary purpose was educational, but encapsulated in storytelling as opposed to lecturing, is a key to what the media might become in the future.

COOKIE MONSTER

When we were children, growing up in the 1960s in suburbia, my sister and I would wake up very early in the morning, sneak down the stairs, open the wooden cabinet doors to the television, and turn it on. We were no more than four or five years old, but having the television hidden away in that piece of furniture and being able to turn it on ourselves made every morning like Christmas morning, the excitement of unwrapping a new package and seeing what was inside.

What was inside most mornings at 5:30 a.m. was generally a television test pattern with the head of an American Indian superimposed upon it, yet we could sit and patiently wait, listening to the test-pattern tone, knowing that a myriad world of entertainment, from *Agriculture U.S.A.* to *Merry Melodies* cartoons, was on its way.

As it turns out, we were not alone. Doubtless, across America, small children were doing exactly the same thing. And that included three-year-old Sarah Morrisett, who was sitting silently staring at the test pattern in her parents' home in Irvington, New York.

Her parents, Lloyd and Mary, generally slept through the early morning video entertainment, as did my own parents in Cedarhurst, New York. However, on one Sunday morning in December 1965, Lloyd Morrisett wandered downstairs and watched as his daughter sat transfixed before the glowing screen.

I have no doubt that our own parents, my sister's and mine, probably also noticed our behavior. I am sure our behavior was no different from that of millions of young children around the world, confronted with this new and very addictive toy in their own homes. My parents, had they seen what we were up to, and I am sure they must have, would simply have retreated back to bed, probably with the comment, "idiots," and that would have been that.

But my parents did not have a PhD in experimental psychology from Yale. Lloyd Morrisett did, and he began to wonder, *What is a child doing watching a station-identification signal, and what does this mean?*[34] A few months later, at a Manhattan dinner party hosted by Tim and Joan Cooney, Morrisett tossed out his observation of his daughter's relationship to the television set as just dinner-party small talk. But it was a comment that found a fertile recipient in Joan Cooney.

Cooney was then working as a producer at WNET/13, the New York local PBS station, the very same place I would get my own start years later. The station had originally been launched as WNDT, but the call letters had been changed to WNET, the NET part standing for National Educational Television. These were still the early days of television, and even earlier days of Public Television. PBS as a network would not be launched until 1970. But as early as 1961, Newt Minnow, then the chairman of the FCC, had dubbed television a "vast wasteland." Everyone instinctively understood that an incredibly powerful and addictive medium was being wasted.

During the course of the dinner party, Morrisett turned to Cooney and asked, "Do you think television could be used to teach young children?" It turned out to be a seminal moment in media history. Cooney was a television producer, a fairly rare breed in those days. But Lloyd Morrisett was not just a Yale psychologist. He had recently been on the receiving end of a $1 million grant from the Carnegie Corporation to find ways to "enhance the educational experience of young children." It was, in a way, a perfect match.

---

34 Michael Davis, *Street Gang: The Complete History of Sesame Street* (New York: Penguin Books, 2009). I am deeply indebted to this excellent book for the story of the Morrisetts.

The upshot of that conversation would be the creation of *Sesame Street*, created by Joan Ganz Cooney and Lloyd Morrisett. *Sesame Street* would go on to become one of the most widely watched and most influential programs in television history. More than eighty-six million Americans had watched *Sesame Street* by 2018, making it the fifteenth most-watched program in American history. But its influence was not just limited to America. The *Sesame Street* format was replicated in more than 140 countries. In the United States alone, *Sesame Street* has won an astonishing 193 Emmy Awards and even 11 Grammys for its original music.

But what makes *Sesame Street* particularly interesting to us is that it was not created to entertain, at least not per se. Rather, it was created as an educational tool—education wrapped in entertainment, just like *The Archers*.

The writer Malcolm Gladwell has said that "*Sesame Street* was built around a single, breakthrough insight: that if you can hold the attention of children, you can educate them." *Sesame Street* holds the attention of children through the use of recognizable and repeatable characters. These are the well-known cast of muppets that Jim Henson invented: Kermit the Frog, Big Bird, Bruno the Trashman, Cookie Monster, and so on. The important point here is repeating characters, along with arcs of story. This is, in effect, little different from what makes any other story work, and we know the educational power of storytelling. In *Sesame Street*, Cooney and Morrisett were able to marry the power of storytelling and character to education to the remarkably addictive nature of television. For children, it turned out to be a winning combination.

SESAME STREET FOR ADULTS

But suppose you could do the same thing for adults. Suppose you could create a kind of *Sesame Street*—repeatable and likeable characters married to engaging arcs of story—for adults. Suppose you could employ the highly addictive and penetrative power of video and television to educate, as opposed to just entertaining or amusing or frightening adults, in search of advertising dollars. Would it be possible? And if so, how would you go about doing that?

The public educational system that we use now was developed in the late nineteenth century, sitting at desks, listening to a lecture, taking an exam. Public education itself was developed more to mold rural children into good factory workers. It had all the architecture of factory work, once you start to look at it. Come in at the right time. Do the tasks you are instructed to do, as you are instructed to do them. Test how well you have learned the lessons. Shifts marked by a bell. Out at three. Just like a factory. Except we don't live in a world of factories any longer, for the most part. We live in a world permeated by the ever-present media. Instead of seeing the media as the distraction from education, perhaps the media is actually the path to education, but done in a very different way, and also not ending at the age of eighteen.

Many people in the educational business will acknowledge that in a world permeated by smartphones, tablets, laptops, and never-ending streams of entertainment, the education business might be ready for a kind of uberdisruption. But most of the solutions they come up with are simply putting video cameras in traditional classrooms and putting current classes online. Maybe the *Sesame Street* model for adults could be more effective.

However, if you are waiting for NBC or CNN or Fox to create a *Sesame Street* for adults, a place where you could really educate people on an ongoing and never-ending basis about things that really mattered, you will be waiting a long time. That is not their mandate. They are in the business of making money for shareholders, and they are quite comfortable with the business model that they have both created and refined over the past seventy years.

That brings us back to your smartphone and the Internet. All the tools you or anyone else needs to create on your own a very different use of this addictive medium are in your hands. Joan Cooney had WNET/13 at her disposal when Lloyd Morrisett approached her. You have your phone, the last gift of Steve Jobs. You don't need NBC or Fox News or CNN or PBS or Google or Facebook to change the world. You can do it yourself. You have the power to change the world.

YOU AND EVERYONE ELSE

At first glance, Pattillo Higgins was not the kind of man who made an impression on you. Nor would you think him the kind of man who would change an industry, let alone the whole world. But that was what he was destined to do.

Born in the small town of Sabine Pass, Texas, to Robert James and Sarah (Raye) Higgins on December 5, 1863, Pattillo had all the earmarks of a lifelong failure. At the age of six his family moved to Beaumont, Texas, where he proceeded to drop out of school after fourth grade. He ended up apprenticing with his father, a gunsmith. As a teenager, he was a loudmouth and a troublemaker. At age seventeen, he got into serious trouble with the law when he began harassing a local black preacher, apparently a regular activity of his. In this case, a local deputy tried to stop him. He pulled out his gun and began to shoot at the deputy. The deputy returned fire, and when it was all over, the deputy was dead. Pattillo had been shot in the arm.

An investigation followed, and Pattillo was cleared on the grounds of self-defense. The wound to his arm, however, grew septic, and the arm had to be amputated. That put an end to his short-lived career as a gunsmith, so he headed off to various logging camps in Texas and Louisiana. Remarkably, he seemed to have found some success as a one-armed logger, and it seemed not to have put a dent into his wild ways.

However, in 1885, his life took a dramatic turn following a Baptist revival meeting. He accepted Jesus into his life, gave up the logging world, and returned to settle down in his native Beaumont, Texas, and take up the life of a respectable businessman. As he explained, "I used to put my trust in pistols . . . now I put my trust in God." The conversion was so complete that Higgins began teaching Sunday School classes. He took the money he had saved in the logging business and founded the Higgins Manufacturing Company, a brick-making and glass-making firm. He also became something of an amateur geologist.

Brick and glass manufacturing required the even-burning heat provided by either gas or oil. At that time, almost all the fuel burned in the United States, and indeed in the world, was coal. Oil was both a rarity and expensive. Much of the world's oil came, in fact, from whales—a difficult, dangerous, and inherently limited source of supply.

In 1859, Edwin Drake had successfully drilled for oil in the ground, what was then called rock oil, in Pennsylvania, with some success. But Drake's oil supplies, and subsequent other eastern oil fields, were also small and limited. It seemed that rock oil would soon run out while coal supplies seemed absolutely limitless.

Just outside of Beaumont, there was a grassy hill called Spindletop. It was a favorite place for church gatherings, and Higgins had been there many times with his church group. Higgins' amateur study of geology led him to believe that there might be oil underneath the Spindletop hill. In geological terms, the structure was called a salt dome—an upwhelling of underground salt pushing through sedimentary rock. No one else believed him or even thought it possible. The conventional wisdom of the time said that Texas, and in fact the entire Gulf Coast region, had no oil. Higgins believed otherwise, and with financial backing from George W. Carroll, whom he knew from his church (and later a few other investors), Higgins proceeded to set up an oil drill and drill down into Spindletop.

After a year of stop-and-start drilling, Higgins had come up dry and was nearly out of money. Industry experts believed Higgins to be a fool. His investors doubted that they would ever see a dime of their money back. But Higgins found new investors and pressed on.

Then, on January 10, 1901, Higgins struck oil. He struck oil in a way that no one until then ever had. At 1,139 feet, his drill bit tapped into a massive reserve of underground oil and gas. The pressure of the gas immediately blew a geyser of oil 150 feet into the air and continued to spew oil out for nine days until it was capped. Spindletop flowed at an estimated rate of one hundred thousand barrels of oil a day. Higgins had not just tapped into the biggest oil find in history, he had also tapped into the future.

Nothing like this had ever been experienced before. A new age had been born—the petroleum age. It was the event that gave birth to Exxon, Mobile, Texaco, and Gulf Oil. It also gave birth to the automobile, the airplane, the world of plastics, pharmaceuticals, fertilizer, mechanized farm equipment, and pretty much everything we take for granted today.

Pattillo Higgins and Spindletop opened the door to a virtually limitless supply of cheap and dependable oil, the fluid that would become the lifeblood of the twentieth century and beyond. Prior to the strike at Spindletop, it was believed that there was very little oil in the ground. The small wells in Pennsylvania and Ohio were so small that, despite the discovery of oil, most automobiles built at that time were either electric or steam. No one thought oil would have much of an influence.

What Spindletop was to oil and the industrial future, the smartphone is to the media future. It was once thought that the only sources of media content were Disney or Comcast or CBS; they were the Ohios and Pennsylvanias of media. But what happens when you turn on and empower the 3.5 billion people who already have smartphones to start making and sharing and broadcasting content that they care about, that has real meaning, and that matters?

When the printing press was unleashed on society some five hundred years ago, it was a tool that vastly reduced the cost and complexity of creating books, Bibles in particular. Bible-creation and book-creation, such as there was, had until then been the exclusive domain of the Church and the monarchy. But it was neither the Church nor the monarchy that embraced the power inherent in the printing press. It was, in fact, average people who would become the printers and publishers of the print revolution. So now, with the digital and video revolution, it will not be NBC nor PBS, for that matter, that will create the *Sesame Street* for adults, but rather you and me. And all we need do is do it.

Even though the printing press was invented in 1440, it is astonishing to realize that the first modern novel, *Robinson Crusoe*, was not published until 1719, some three hundred and fifty years after Gutenberg first published his Bible. Once again, the technology was always there. What was missing was the architecture of the novel.

Up until now, the media has been used to maximize audiences to sell advertising. This worked quite well when the control of the media was in a very few hands. But, as Gutenberg and his contemporaries could only see the printing press as a device for producing and distributing religious tracts because that is all they had ever known, so too can our contemporaries often only see the web and the smartphone as a device for doing more of the same. It could be used for that, but it could also be used for something far more interesting and perhaps far more beneficial for society as a whole.

In the world of media, you have a captive audience—in fact, an addicted audience. They are not giving up their addiction. If you craft your content correctly, you can be as compelling as any TV show or online video. But now you have the opportunity to marry that medium, that technology, and that addiction not just to garnering more eyeballs or more "likes"; you can use the technology and the medium to educate people, an ongoing education that lasts a lifetime.

Of course, there are endless videos on YouTube for "education," but they are as unwatchable now as the droning teacher in the classroom. But wrap the education in the basic principles of storytelling, marry it to the addictive nature of our visual medium, and you have a revolution.

Human beings are like sponges. They absorb knowledge. Up until now, the media has been teaching, and people have been absorbing, but the lessons they have been absorbing have been largely detrimental—fear, anxiety, terror, depression. But it could be different, better, enlightening, elevating. CNN and CBS won't do it. But you can.

13

The Idiot's Guide to Taking Control of the Media and Your Life

OR
YOU AREN'T GOING TO CHANGE THE WORLD WITH SELFIES

Understanding both the addictive nature of video/television and the enormous manipulative control that it has had over your life up until now, you may be motivated to simply cut the cord, pull the plug, throw your phone out the window, and never go near anything except a physical book again. That would be an intelligent response, but you know and I know that this is not going to happen.

It is not going to happen because video, as it turns out, is far more addictive than cigarettes or heroin and cigarettes and heroin are not woven into the fabric of society. The economy, politics, day-to-day transactions, and simply being a functional part of society are not predicated on smoking or shooting up; they are, however, bound up tightly with video/television content. What we watch is who we are.

TAKE CONTROL OF THE MEDIA, AND YOUR LIFE

Unable to break the addiction, you must then do the next best thing, and that is to take control of the medium yourself, to master it and to use it for your own ends and, perhaps, to create a better society. In the last chapter of this book, I am going to deal with ways in which the enormous power of media can perhaps be harnessed for the common good (as opposed to scaring you or trying to convince you to buy more Pringles). The potential for this is enormous. An addictive media is far too powerful to be left in the hands of corporations.

The only way to effect any change, then, is not to abandon the media but to learn how it works and to take personal control of it. The tools to do this are already in your hands; the question now is, how best to use them.

Posting selfies with cat's ears on Instagram, even with Boomerang, or videos of your vacation to the Bahamas is not going to change the world, nor does it come even close to getting your hands on the levers of power that the media, properly understood, has the potential to exercise. You are not even scratching the surface. Neither will posting these banalities, entertaining though they may be, free you from the extraordinary control over your life that the media currently exercises. You must do more. You must become literate. You must get your hands dirty. You must take control.

This, of course, is exactly what Martin Luther did with the printing press. He seized the new technology with both hands, learned, and used it to change the world. Martin Luther did not use the technology to print selfies.

But simply seizing the technology is not enough. Everyone has a smartphone today, and everyone uses it. It is a tool with remarkable potential to effect change, but to do this, you must understand more than simply how to push the button and record video. You must understand why some video works and, understanding that, apply those lessons to things that really matter.

In other words, you need to understand the secrets of exactly how the medium is used to manipulate your emotions, your thinking, your purchasing decisions, and even your view of yourself. That's the bad news. The good news is that the secret is not really so secret. Like any best-kept secret, it is actually out in the open, the last place anyone would look for it. All you have to do is learn to watch in a different way. The best media teachers in the world are right there in front of you every day, and for $12 or so, you can study with them for as long as you like.

DENZEL WASHINGTON AND ME

When I was in my early twenties, I got my first job in television. Those were the days before the Internet, before YouTube, and even before cable. If you wanted to work in TV, you got yourself a job with a TV station or network, or if you were really ready to scrape the bottom of the barrel, you got yourself a job with PBS. And below the bottom of the barrel, you got yourself a job with a local public-television station.

And thus it was that my very first job was as a production secretary at WNET/13, the local public television station in New York. And, if that was not bad enough, I was assigned to their Newark, New Jersey, offices. WNET/13, New York's public television station, was not actually licensed in the state of New York. It was licensed in the state of New Jersey, but as no one wants to live or work in New Jersey unless they have to, all of the offices and studios were based in midtown Manhattan, in the Henry Hudson Hotel on West 58th Street. That is, all of them except for a kind of Potemkin studio in the Gateway Office Complex in downtown Newark, next to the train station. The Newark office was there to protect the license, in case some local New Jersey types tried to challenge the license and actually put the station in Jersey. This was where I worked.

So, while the New York offices were producing things like *Live From Lincoln Center* and *Masterpiece Theater*, things that won awards, the Newark office was producing things like *Mainstream*, a New Jersey public-affairs talk show that involved two guests, two swivel chairs, a New Jersey state flag, and a potted palm. *Mainstream* aired at 6:00 a.m. on Saturdays. The ratings were small. I tried to convince my mother to watch, but she said it was too early. *Mainstream* was designed to protect the New Jersey license. "Of course we are covering New Jersey! Don't you watch *Mainstream*??" What legislator would dare to say no?

Thus it was that my very first job in the television industry was as a production secretary on *Mainstream*. My job was to cut the bagels and put out the coffee for the guests. That was about it. But then again, producing a half-hour talk show once a week does not exactly require a massive team, nor much effort on anyone's behalf.

In any event, down the hall was the science show called *Innovation*, which aired after ours. One day, there was a big general meeting for all WNET/13 employees, to be held in the New York offices at the Henry Hudson Hotel. It was the 1984 elections, and there was going to be a Democratic primary in New York. WNET/13 had decided that it was going to do a special live broadcast on the night of the primary. This was a very big deal for WNET/13, particularly the live part. This also showed the station's commitment to both politics and to public affairs.

This was going to be a special, so there was no staff in place to produce it. As such, there was to be a general meeting of all employees who were interested in working on the special live primary election event. They weren't paying anything extra, but it would look good on a resume. Everyone was invited to Room 101 at 5:00 p.m. on Tuesday at the Henry Hudson Hotel.

I went to the meeting. There must have been about a hundred people in the room. Since I worked in the Newark office, I didn't know anyone, and no one knew me. Most people didn't even know that there was a Newark office. Joan Konner, who was both the vice president of WNET/13 and the executive producer of the New York primary special got up and explained how all this would work. The candidates would actually be in the studio, on swivel chairs. There would be three videotape pieces that would air before each candidate spoke. These would be one-to-two-minute miniprofiles of the candidate's day in New York. It was a daring and bold exercise in the free electronic press, at least for Channel 13.

After the explanation was over, Joan asked anyone who wanted to produce one of the three pieces to raise their hands. Now, the only thing I had ever produced in my life was coffee and bagels, but Peggy Girshman, a producer from the science show down the hall, who had more or less dragged me to the meeting, elbowed me and said, "Raise your hand," so I did. And before I knew it, I had been tapped to be a producer of one of the three videotape candidate profiles that would air that night.

In 1984, there were three Democratic candidates running for the nomination: Walter Mondale, Gary Hart, and Jesse Jackson. I got assigned Jackson. I was supposed to go to the Hilton Hotel the next night, just prior to the live New York State primary election special airing, and shoot and edit a one-minute profile of Jackson and his rally at the Hilton. Not really knowing what I was doing at all, but thinking, *How hard can this be?* I arrived at the WNET/13 offices at the Henry Hudson Hotel around 3:00 p.m. on the day of the special broadcast and went to the field shop. The field shop was where the union video crews could be found, buried amidst their piles of cameras, bags, tripods, lights, magazines, old newspapers, half-eaten sandwiches, and God only knows what other kinds of stuff. I had been assigned to work with the team of Dale Vennes and Nick Pavichavich. Dale was the cameraman; Nick was the sound guy.

I had hoped that, as they were an experienced crew, they would help me out. As soon as they saw me, they knew they were in for a fun time.

"Oh, Mister Producer," they said, "are you new here?"

I said I was, and as they gathered up the gear, they asked me, "Oh, Mister Producer, what kind of lenses do you want us to bring with us?

If you had asked me sesame or poppy-seed bagel, I would have known. When it came to lenses, I had not a clue. But I did not want to look stupid on my first job, so I said, "The standard ones." That seemed reasonable.

"Oh, Mister Producer, what kind of microphones should we pack?"

"The good ones."

Now, they had probably done this kind of thing a thousand times. I had never done it before. They could have gone in without me and come back with all the elements of a perfect story, but that was not what they did. Instead, when we arrived at the Hilton, they only shot exactly the shots I told them to get and nothing more. And who knew what shots to get? Certainly not I!

We stayed far too long at the Hilton, and now, with the clock running, I had to get back to the Henry Hudson Hotel and edit together this pile of junk that I had accumulated at the Jackson rally. That was not going to be easy, and again, I had absolutely no idea what I was doing.

When I got back to West 58th Street, I was assigned to work with a very experienced editor named Freddy Rodriguez. In those days, edit suites looked like NASA's control room for a flight to the moon. There were hundreds of dials and meters and switches. Freddy sat in the center of the master control; I sat behind him. I was supposed to tell him what to do. I had absolutely no idea. I was just making it up as I went along. Each time I would say what shot to put in, Freddy would turn around to me and say, "Are you sure you want to do that, man?" Watching the clock, and seeing it get later and later, I just said, "Put it in."

Soon, the live show started downstairs in Studio 1A, the biggest studio at WNET/13. I was still feverishly working away with Freddy up in the edit suite on the fifth floor when the phone rang. It was Joan Konner. "Where is the goddamned Jackson piece?" she wanted to know.

"It's coming!" I said. She slammed down the phone.

Freddy worked away at a fever pitch, finishing up the story. Soon, he handed me the three-quarter-inch U-matic tape cassette, and I ran down the five flights of stairs, clutching my tape, not wishing to wait even for the elevator. The live show was already halfway done.

I ran down the corridor and burst into the control room for Studio 1A. Joan Konner was seated in the main chair, a giant leather swiveling lounger that looked like the command seat for Captain Kirk on the Starship Enterprise. The rest of the room was swarming with people—technicians, sound engineers, lighting people, and pretty much every major executive from WNET/13.

Before us was a giant plexiglass wall, and on the other side, you could see the three candidates and the host. The cameras were moving, the images filled all the monitors in the control room. The show was deep in process.

"Here's the Jackson piece," I said, out of breath, as I handed Joan Konner the tape. She snatched it from my hands, spun her chair around thirty degrees, shoved the tape into a playback deck next to her, hit play, and started to screen my one-minute masterpiece. About halfway through it, she hit the eject button, withdrew the cassette, stood up, and threw the tape against the plexiglass wall before her. It hit with a resounding crash and fell to the floor. The room, needless to say, went silent.

"This is the biggest piece of shit I have ever seen," she yelled to all assembled, me included. Then, she pointed her finger to the door. "Get out! You'll never work in this industry again!"

This, was, needless to say, not a great career move nor moment for me. I slunk out of the control room, with all eyes of every executive in the building upon me. I was fairly certain that my nascent career in television was already over before it had hardly begun. In the next few weeks, I busied myself keeping my head down, cutting bagels, and reading up on law schools. Fortunately, as I was back in Newark, I was able to avoid what would have been the searing looks from pretty much everyone who worked for Channel 13.

However, television is nothing if not ephemeral. About six months later, it was announced that the Department of Defense was going to close Fort Dix, a US Army base, which, as it happens, was in New Jersey. WNET/13 decided that, most likely in their never-ending defense of their license, they would do a special program on the closing of Fort Dix.

Once again, the word went out that anyone who was interested in working on the Fort Dix closing special should come to Room 101 on Friday at 3:00 p.m. Prodded once again by Peggy Girshman, we both went over to the Henry Hudson Hotel and filed into Room 101, where once again, there were about a hundred people.

"Who wants to work on the program, raise your hand," some amorphous voice from the front yelled out. Peggy elbowed me in the gut, so I raised my hand. Some woman clutching a clipboard came over to us. I didn't know her, and she didn't know me.

"Have you ever produced anything before?" she asked.

"The live New York primary election special," I said, which was entirely true.

"Good," she said. "Come to the crew shop on Monday morning and pick up a crew. You're in."

Now, as it happens, the movie *A Soldier's Story* was just opening that weekend. It was the Denzel Washington breakout movie. On Saturday morning, I bought a ticket and watched the movie. It was, after all, about the army. In fact, I watched *A Soldier's Story* twelve times that weekend, over and over, trying to understand how it was made, carefully taking notes on how it was shot and put together:

Close up on the flag.

Cut to the trumpeter.

Close up on the shoes.

Wide shot of the parade grounds.

Close up on the gun.

On Monday morning, I appeared at the WNET/13 field shop at 9:00 a.m. to meet my assigned crew. It was Dale Vennes and Nick Pavechivich.

"Hey, Mister Producer!" they greeted me. "You still got a job?" I affirmed I had.

We piled into the WNET/13 van and headed off to Fort Dix. When we got there, Dale and Nick once again asked what kind of shots I wanted. This time I was ready:

Close up on the flag.

Cut to the trumpeter.

Close up on the shoes.

Wide shot of the parade grounds.

Close up on the gun.

When we got back to WNET/13, I was assigned an editor to cut the segment. I got Freddie Rodriguez. He was both surprised and delighted to see me. "Hey man, you still workin' here?" I affirmed I was.

As we started the edit, I told Freddie how I wanted the segment cut:

Close up on the flag.

Cut to the trumpeter.

Close up on the shoes.

Wide shot of the parade grounds.

Close up on the gun.

What I did was essentially replicate *A Soldier's Story*, except at Fort Dix. When the segment aired on local Channel 13, it was so widely praised that the *MacNeil/Lehrer Newshour* picked it up for national distribution. "I don't know what you are doing wasting your time as a production secretary over there in Newark," Joan Konner said to me. "You should be producing."

And that is how I became a television producer. I would go on to win many awards, to produce for CBS and later to produce more than eight thousand hours for cable. But I did not learn how to do it in journalism school or film school. I learned how to make compelling television from watching movies, the way some people learn English from watching TV.

Now, you see what I just did? I told you a story. I told it to you exactly as Joseph Campbell explains a good story should be told. An unknown rises to a challenge that is far beyond him or her, tries and fails, then tries and succeeds—often with external help—and goes on to great success. In my story, I am the character, and in the end, I am the hero. OK, I am the hero of Channel 13, which is not exactly much of an achievement, but the format remains the same.

SPIDERMAN AND ME

It's all there. The best filmmakers in the world are showing you, for $12, every secret of telling a great story in this medium. When you go to the movies, you aren't paying $12 to see a movie; you are paying $12 for the best education in visual storytelling money can buy. Who better to learn from than Spielberg, Scorsese, Hitchcock, Kubrick? *A Soldier's Story* was directed by Norman Jewison. He also directed such hits as *Agnes of God, The Thomas Crown Affair, The Russians Are Coming! The Russians Are Coming! Fiddler on the Roof*, and others.

Now, notice something else. The story teaches. It is a compelling story, but it also educates. The point of the story, of telling you the story, is to convey a lesson to you. This is the real power of the medium. Edward R. Murrow was wrong. The machine can teach. It is not a question of if it can teach—it can. It is now a question of what it has been teaching.

As you have seen in prior chapters, this addictive medium has, until now, been used, for the most part, just to sell you stuff, and mostly stuff you don't even want. The media companies are masters of using classic storytelling techniques, married to this very addictive medium, to hook you on their drug and then use it to sell you and your wallet to the highest bidder, with absolutely no regard for your mental or spiritual health, or the good of society as a whole, for that matter.

But now you have the very same tools and access that they have. And now you can turn the tables on them and learn to use those tools both to liberate yourself and to make a much better society for all of us.

When Gutenberg first invented the printing press in 1440, Europe was effectively illiterate. What, after all, was the point of knowing how to read and write when the average person probably never came across a hand-written book? Even the great King Charlemagne struggled to write his own name. The advent of the printing press meant that for the first time in human history, the extraordinary power of print was now open to anyone. But to use it, people first had to learn how to read and, more significant, how to write. Otherwise, what was the point of being able to publish?

Just throwing together a bunch of typeface letters in a press, applying ink, and printing whatever came out in 1440 was the equivalent of posting cat videos on YouTube or making selfies for Instagram: amusing but fundamentally pointless. In order to use the printing press to change the world, it was necessary first to learn how to write in a compelling way, how to marry language to the printed word to communicate ideas, to inform and often to change people's minds. That was the power of print. It's not just in the writing, it's in the storytelling. You remember the power of stories.

If video is now going to become our primary means of communicating ideas, and it seems that it is well on its way to doing that, then we must all learn the tools of video literacy so that we may be able to use this new technology to the fullest. This goes far beyond simply pointing a camera at something and hitting record, or appending funny hats or cat ears to yourself and then boomeranging the video and putting it on Instagram. You are not going to change the world with that.

Video is a language. It has the potential to be a remarkably powerful language, perhaps even more powerful than the printed word.

The written word, when read, has to be perpetually decoded. There's a buffering process that takes place in your brain—letters to words to ideas. But video is direct to your brain and to your emotions. It is visceral; it is raw. But to use video to its fullest, you must first be literate in its grammar, much as you must be in the written word. Simply turning on a phone or a camera and either speaking into it or pointing it at something interesting to capture is using only one percent of what this new language can do.

More often than not, the true potential of a creative technology may lay dormant for many years. There is a good deal of debate in literary circles about whether *Robinson Crusoe* (1719) or *Moll Flanders* (1722) was the first English language novel ever published. *Don Quixote* (1605) lays claim to the first modern novel overall. What makes this of interest to us is that any of these (or any of the other competitors for the title) long postdate the invention of the printing press in 1440. Once again, the technology was there, as was the human tradition of great storytelling. Yet the marriage of these two would take several hundred years.

There is no question that video/television/film represent an extraordinary medium for telling stories and teaching and transmitting information in a very direct and powerful way. But if we are going to learn to use this medium ourselves, if we are to seize control of the medium, or at least a part of it, then it behooves us to also learn how to use it to tell stories that have the power to make people think or to change the way that they think. Why should this extraordinary power be solely resident with corporations only driven by profit and sales? Can't this be used for more and better purposes?

IN A WAY, YOU ALREADY KNOW WHAT TO DO

In his book *Outliers*, author Malcolm Gladwell popularized the idea that anyone could become an expert in anything if only he or she was to practice it for ten thousand hours. Play chess for ten thousand hours, and you can become a world-class chess player; practice the piano for ten thousand hours, and you can become a concert pianist; play tennis for ten thousand hours, and you can head for Wimbledon.

The so-called ten-thousand-hours rule may have been popularized by Gladwell, but it was first promulgated in a 1993 paper written by Anders Ericsson, a professor at the University of Colorado,
called "The Role of Deliberate Practice in the Acquisition of Expert
Performance." Ericsson had based his paper on work done in Berlin
by a group of psychologists who traced the career path of student
violinists.

By the time you have reached the age of twenty-one, you have
already seen thirty thousand hours of movies, video, and TV shows.
That means that by twenty-one, you are already three times the world
expert in filmmaking and video. You innately already know how to
express yourself in video; you just don't know yet how much you
already know.

It was the same with *A Soldier's Story*. Had I never seen a film
before in my life, I probably would have had a hard time extracting
from Jewison the secrets to capturing a viewer and holding his or her
attention. I would have been so thoroughly overwhelmed by the sheer
visual and auditory display going on before me. But I had already
seen hundreds if not thousands of movies, and as I watched *A Soldier's
Story* over and over, again and again, I could see that there were pat-
terns, not just in Jewison's movie, but in fact, as I thought about it,
and later saw, in every movie and TV show I had ever seen. They were
all the same.

What makes great storytelling work in movies is two things: a character and an arc of story. If you want to take control of this very powerful medium, if you want to stop being a victim and instead not just free
yourself, but actually empower yourself, then it is absolutely essential
that you learn the grammar of communicating in images and sound.

In Chapter 1 of this book, we discussed the way that, in 1969,
the media bifurcated between reality (the moon landing) and fiction
(*Star Trek*), and that fiction won. The real Space Race, as it turns
out, was not between the Americans and the Russians; it was between
the Apollo astronauts and the crew of the Enterprise. The fictional
Enterprise crew won because what they were offering was more exciting. It had all the elements of great storytelling—characters you could

relate to and compelling arcs of stories. It also had something else—characters that kept coming back, episode after episode.

CHARACTER IS DESTINY

Heraclitus (born 535 BCE), the Greek philosopher, wrote, "Character is destiny." In any kind of storytelling, it is the character that is paramount. We are attracted to any story because we are invested in the character.

Casablanca is a great film because we come to care about what happens to Ilsa and Rick. In *Harry Potter* we care about what happens to Harry and his friends. We experience the book through their eyes, not our own. Character is destiny. In the age of media, character, as it happens, is also money. If you have a message to deliver, no matter what it is, you must drive it with and through a character.

BEFORE THE KARDASHIANS, THERE WAS COTTON

People have grown cotton for thousands of years. The oldest cultivated cotton plants date back to 6000 BCE, found in Huaca Prieta, Peru. Cotton has been found in India, Egypt, the Middle East, and China, all going back thousands of years. But for almost all of its history, cotton was an incredibly labor-intensive plant and fabric with which to work.

Separating the seeds from the cotton fibers in the cotton boll was difficult work that could only be done by hand or by primitive machine. This made cotton, while incredibly important for local and domestic production, an expensive product with which to work. Then, in 1794, Eli Whitney was granted a patent for his new invention, the cotton gin. It was a machine that was capable of separating cotton seeds from fibers, and could process up to fifty pounds of cotton a day. Prior to the cotton gin, cotton had been so difficult to clean by hand, so labor-intensive and time-consuming, that growing cotton was not really a profitable venture. It took an entire day of grueling manual labor to separate the seeds from the fibers in a single pound of cotton. The cotton gin changed everything.

In the United States, the cotton industry exploded, along with, terribly, the slaves that were necessary to both harvest the vast quantities of cotton it processed and run the machines. Enormous, almost unthinkable fortunes were made, and cities such as Galveston, New Orleans, Charleston, and Mobile grew rich on cotton exports. By 1860, the American South was providing two-thirds of the world's cotton supplies, and eighty percent of the critical British market. Cotton, thanks to the cotton gin, was the backbone of the American economy.

The ability to mass-produce vast amounts of cotton also became the great driver of the Industrial Revolution in Britain, where the cotton was spun into cloth that was then sold around the world. By 1850, Britain had 1.8 percent of the world's population, occupied an area 0.16 percent of the world's land mass, and yet produced half of the cotton textiles in the world. The wealth of the British Empire was a direct result of the invention of the cotton gin.

The simple cotton boll was so transformed, became so valuable in its own right, that as the South began to lose the Civil War and it could no longer back its paper currency by gold, it began to back its paper currency with cotton. Cotton bonds became the most valuable source of money the Confederacy had. The extraordinary transformation in the perceived value of cotton from next to nothing to the backing for currency was driven by a simple piece of technology and nothing more. It was the tech that took the value of a cotton boll from interesting plant to the foundation of a global economic boom. The boll remained unchanged; it was its marriage to the tech that made it suddenly so very valuable. This is the power of a new technology, properly applied. And the lowly cotton boll is not an anomaly. This story has lessons for us today.

MAKING A BILLION DOLLARS FROM AIR

In December 1993, Joanne Arantes was in serious trouble. Born in England to middle-class parents, she had graduated from the University of Exeter and had gone off to Portugal to teach English as a second language. While there, she had met and married a Portuguese television journalist named Jorge Arantes, and they had had one child together.

But by 1993, the marriage had gone badly, and Arantes took her daughter and moved back to the United Kingdom, renting a small flat in Edinburgh, Scotland, so she could live near her sister. Jobless, with a dependent child, living on state benefits, she felt her life a complete failure. She slipped into a clinical depression and at one point considered suicide. She spent most of her days sitting in a nearby café writing a novel that she had been thinking about for some time. The novel was *Harry Potter*, which she wrote under her unmarried name, J. K. Rowling. Harry, of course, went on to worldwide fame and astonishing success. Today, the *Harry Potter* franchise has generated more than $25 billion dollars.

Now, here is the interesting point about all of this. With the stroke of a pen, literally, J. K. Rowling created, from nothing but a concept, a single character worth $25 billion. And who is Harry Potter? Harry is a concept, a character, an image—an imaginary image to be sure, but an image. Harry Potter has less presence to him, less physical reality, than even a cotton boll. In effect, this image, created from nothing and made of nothing, is more valuable than air and more powerful. What is Harry Potter that he has so much value? You cannot live in Harry Potter, you cannot eat it, and you cannot wear it (which at least you can do with cotton). Harry Potter does not exist, has never existed, and will never exist, but look at the value that we put on Harry Potter.

Each new technology brings enormous value to things that prior to that technology had little to no worth. Before the invention of the cotton gin, there was little value to cotton; post–cotton gin, it fuels the Industrial Revolution and the British Empire. Prior to the media revolution, fictional characters in novels were about as interesting and as valuable as seedy cotton bolls. They existed, they were nice, but no one ever appended a value of $25 billion to Mr. Rochester or Huck Finn. Books were great. Charles Dickens made a nice living from his own fictional characters, but he was no J. K. Rowling. It would take the electronic media machine and addictive moving images to do that.

The cotton gin created a multiplier value for cotton bolls. The media machinery of the late twentieth century created a multiplier for characters. There had always been characters since the time of Homer. Characters are the mainstay of storytelling, and we know what power storytelling has in human history, but it had never been so valuable. A character from a story had never before been such an incredible commodity in its own right. J. K. Rowling effectively created $25 billion from the air, from whole cloth (as the cotton people might have said). It was a remarkable achievement, and it bore testimony to the extraordinary value that we now place on characters, fictional or real.

And it was not just Harry Potter. Spiderman, a comic-book character created by Stan Lee, also completely fictional, has generated more than $6 billion, and that is just in movie receipts. The total gross revenue from the franchise puts him close to Harry Potter in value. Add them to the legion of fictional billionaires that includes Darth Vader, Luke Skywalker, Iron Man, Superman, Batman, and God only knows how many others. All are incredibly valuable, all created from nothing.

In a sense, Harry Potter is little different from Kylie Jenner, net worth $1 billion—a cartoon character in her own right, multiplied by Instagram. This is also little different from the value attached to the *Real Housewives,* the reality TV stars, the winners of *Love Island,* and so many other talentless nonentities who suddenly find their media-driven fame bankable. This is what fuels the endless drive to become an "influencer" on Instagram.

As a society, we place enormous value on our characters. Characters are for us what cotton bolls were to the Industrial Revolution—they are the raw material from which we spin and weave and craft the fabric of a media-driven world (as opposed to a manufacturing-driven world). The nineteenth-century world of the industrial revolution was a world made of things. Our world, in the twenty-first century, is increasingly a world made of characters.

Thus, if you are going to embrace this medium, if you want to control it and own it and use it for your own ends, and hopefully something much better than selling yet more junk, it is imperative that you understand what makes it work. And the core of what makes it work is characters. It is an entirely character-driven medium. Thus, if you want to use it to educate or convince people, you must base that content on a character, not a concept or an idea or date or information. Michael Eisner, as it turns out, was wrong. Content is not king. Characters are king.

ARC OF STORY

Cotton thread is good, but cotton does not really achieve its true value until it is woven into a fabric. The same holds true for characters. A character is interesting, but it only achieves its true maximum value if you can weave it into a story that people want to watch, stay with, and absorb. This is what we call arc of story. The character plus arc of story is what gives you entrée into people's minds and hearts. You may use this to sell Cialis, but you may also use this to sell ideas.

In 1988, I was working as a producer for the CBS News program *Sunday Morning*, which is still on the air and still looks the way it did more than thirty years ago. I found working with a crew frustrating, so I quit, bought a small video camera, and wanted to see if I could make television journalism in a completely different way—on my own. Today, a lot of people shoot their own video stories, but in 1988, this was a very radical and revolutionary idea. I bought myself a small video camera and went to live in Jabalia, a Palestinian refugee camp in Gaza. I moved in with a family and shot video every day. As it happened, I was also there during the first intifada, the Palestinian uprising against Israeli rule. Think of this as good timing.

I spent a month with the family, and shot a very personal story, highly character-driven, of what life was like in Gaza then. I came back to the United States and sold my story to what was the *The MacNeil/Lehrer Newshour* on PBS, another news show that is still on the air, and still pretty much the same, save for the name change. MacNeil/Lehrer paid me $50,000 for my stories, which was pretty good pay for one month's work, particularly in 1988.

Shortly thereafter, I was contacted by a Swedish billionaire named Jan Stenbeck. He was starting the first commercial TV networks in Scandinavia. He understood the economics of what I had done. I had gotten rid of the camera person, the sound person, the producer, and the editor. He flew me to Stockholm and asked me a seminal, life-changing question: "Can you teach other people to do this?"

I told him what I have told the more than fifty thousand people I have taught to do this since then: "Any idiot can do this. All you have to do is follow some basic rules." Working with Stenbeck, we started to build TV stations in Scandinavia based on this concept, which I called video journalism. My idea was to marry the power of great print and photo journalism with video—personalized, highly character-driven, arcs of story.

The project in Scandinavia proved so successful that I was soon contacted by Paul Sagan, who was just starting a twenty-four-hour news channel in New York for Time/Warner called NY1. I met with Sagan and his staff in New York, and Sagan asked me how many video journalists I would have in NY1.

"All of them," I replied.

"And how many crew?" he asked.

"None."

Paul Sagan had courage. He committed one hundred percent to the VJ concept, which was then a total unknown.

I went on to design and help build VJ-driven TV stations around the world. A lot of media companies were attracted to it for the cost savings, but few were interested in the potential it offered for a different kind of journalism. Ironically, some thirty years after the launch of NY1 (which turned out to be a great success), Lisa and I got a call from Spectrum, the company that had purchased Time/Warner's cable operation. The people there were thinking about building a twenty-four-hour local TV news channel in LA to be called Spectrum1. Were we interested in working with them on it?

We were, and we were fortunate to find that the people running Spectrum1 were as open to embracing new ideas as Paul Sagan had been some thirty years prior. We trained the NY1 MMJs (multi-media journalists) to shoot, edit, and produce stories that were a fusion of outstanding journalism and great drama-driven storytelling. The result was an entirely new kind of television news and journalism that has proven to be a massive success. It is the invention of an entirely new way of using storytelling and the medium of video to communicate ideas and information in a compelling way that works. It is character-driven and relies on arcs of story, but it deals in real stories and the lives of real people. And, it is done by individuals.

What NY1 did, what Spectrum1 is doing now, you can do as well.

TODAY, WE ALL ARE JOURNALISTS

The tools to use and manipulate video/television/electronic media are in your hands. You may use them any way you wish, just as the printing press was a tool that could be used in many different ways aside from making Bibles.

Until now, that enormous power and that highly addictive medium has been in the hands of only a very few massive media companies. Their only interest was and is to sell you things. This skewed the content and, as a result, it skewed the way you and the rest of the world saw events and saw yourself.

In an ideal world, I would tell you to throw away your phones, cut your cable, smash your TV sets, and pick up a book and start reading, or just go out for a long walk in the countryside and never look at another screen again. This might be the best advice I could offer, but let's be honest. It isn't going to happen, and it is not realistic. Our addiction to video and TV is here to stay. The technology is now so tightly woven into our culture that the call to burn the media companies to the ground, appealing as it might sound, is just not realistic.

In the nineteenth century, the Industrial Revolution brought sweeping changes to Britain and British society. Small cottage industries and indeed the whole foundation of the village and rural economy were swept away as millions flooded into cities to work in the newly built factories. A way of life was being destroyed. There was scattered resistance and rebellions, but in the end, the Luddites and their ilk could not hold back the overwhelming force of change. Neither can we.

Instead, what we can do is seize control of the media world, or at least a part of it; make ourselves literate in how it works; and then use those skills and tools to drive the common good. It is possible.

One of the unfortunate side effects of our addiction to visual media is that it engenders a sense of passivity. Watching, as you know,

218 DON'T WATCH THIS!

is passive. The more you watch, the more passive you become. Today, we are faced with serious problems that the media world simply seems incapable of dealing with. And why should it? Its job is to make money from the medium, not to repair society.

But you can make a difference. You can change the world. The power of the media machine is astonishing, but it is in the hands of a very few corporations. It can be in your hands. And it should be.

The media companies won't use their power to change the world. But you can. Those days are over. An enormous opportunity is now before you. Take it. Carpe medium.

Acknowledgments

I am, first, foremost, and always, deeply indebted to my lovely wife, Lisa, who saved my life and restored and rebuilt our media companies and took them to enormous success. She also gave me the endless support and encouragement necessary to make this book possible.

I had been trying to write this and get it published for years, and wrote to innumerable agents, all of whom wrote back and told me this was a great idea, but . . . with a long list of buts. About to throw in the towel, I was remarkably fortunate to connect, on LinkedIn no less, with Alicia Brooks, who took to the book with a vengeance and succeeded in getting it published.

After many rejections by publishers, also writing, "Great idea, but . . ." I told Alicia just to get me a face to face meeting with a publisher. Here, I was most fortunate that Tony Lyons of Skyhorse agreed to meet with me. I am, needless to say, deeply indebted to him for agreeing to publish the book, and his support and confidence have been critical to its success.

Tony assigned Rebecca Shoenthal as my editor, and she has brought enormous insight into the writing and the structure of the book, and she contributed greatly to its success. I am most appreciative of all the time and effort she has put in on this, and how immediately responsive she has been to even the stupidest of questions that

I might ask, which are many. I'd also like to thank Jen Garth for her fantastic copyediting.

I also want to thank Mark Bittman, author of the best-selling cookbooks in America, who gave me enormous insight into how the food business in America works. There are, as it turns out, great parallels between food and media, probably the two greatest items of consumption in this country, if not the world.

Special thanks to Charlotte Frank, who at the age of ninety-eight (!) read endless iterations of this and was particularly insightful in helping shape the narrative.